Moral Energy in America

Energy Humanities
Dominic Boyer and Imre Szeman, Series Editors

Moral Energy in America

From the Progressive Era to the Atomic Bomb

REBECCA K. WRIGHT

Johns Hopkins University Press
Baltimore

© 2025 Johns Hopkins University Press
All rights reserved. Published 2025
Printed in the United States of America on acid-free paper
2 4 6 8 9 7 5 3 1

Johns Hopkins University Press
2715 North Charles Street
Baltimore, Maryland 21218
www.press.jhu.edu

Library of Congress Cataloging-in-Publication Data

Names: Wright, Rebecca K., author.
Title: Moral energy in America : from the progressive era
to the atomic bomb / Rebecca K. Wright.
Description: Baltimore : Johns Hopkins University Press, 2025. | Series:
Energy humanities | Includes bibliographical references and index.
Identifiers: LCCN 2024035204 | ISBN 9781421451411 (hardcover) |
ISBN 9781421451428 (ebook)
Subjects: LCSH: Power resources—Social aspects—United States—
History—20th century. | Energy industries—Social aspects—
United States—History—20th century. | Energy consumption—Social
aspects—United States—History—20th century.
Classification: LCC HD9502.U5 W75 2025 |
DDC 333.790973—dc23/eng/20241228
LC record available at https://lccn.loc.gov/2024035204

A catalog record for this book is available from the British Library.

Special discounts are available for bulk purchases of this book. For more information, please contact Special Sales at specialsales@jh.edu.

CONTENTS

List of Illustrations vii
Acknowledgments ix

Introduction 1

1908–1919

1 Energizing the Will: *Progressive Energies* 23

1920–1943

2 Energizing Americanness: *Climatic Energy and Race* 55
3 Energizing Culture: *Psychic Energy and Primitive Waterways* 81
4 Erginette's Beauty: *Energizing the Economy* 107

1939–1951

5 Energizing the World: *Energetic Ethics for an Atomic Age* 139

Epilogue: *MEOW* 171

Notes 177
Bibliography 221
Index 251

ILLUSTRATIONS

Advertisement for the book *Power of Will* by Frank Channing Haddock	24
Illustration of positive heliotropism in polyps of Eudendrium and marine worms (Spirographie)	28
Electric dog invented by John Hays Hammond Jr., and the internal mechanism of the electric dog	31
Illustration mapping the similarity of the tree rings in 1851 across 750 miles of the country	59
Fluctuation of the universal historical process across the globe, between the fifth century and the nineteenth century	62
Clock of cold droughts and civil war	65
Distribution of international and civil war battles occurring in all known parts of the world from 1750 to 1943	66
Map charting levels of human energy against climate	70
City populations and growth mapped against temperature	72
Illustration of Giant Power	82
Illustration of the associative network of the mind	93
Illustration of Appalachian Power	97
Illustration of industrial flow across the United States, and the "sphere of origin" in the Somerset Valley, Upper Deerfield River	98
Illustration of Erginette	108
Illustration of how Technocracy would usher in an era of plenty	111
Illustration of a broken Grecian bust representing Arnold Toynbee's theory about the decline of civilization	147
Diagrams of the different cultural states	159

ACKNOWLEDGMENTS

This book was mostly written in libraries across Bloomsbury, when I was a PhD student at the London Consortium, Birbeck College, University of London, between 2011 and 2015. I am grateful to have been part of the education experiment that was the London Consortium, which exposed me to different ways of thinking and gave me the creative freedom to pursue this project. My supervisors, Steven Connor and Peder Anker, shaped the project and supported my early development as an academic. My examiners, David Nye and Sam Halliday, provided important feedback on the manuscript and allowed me to view the project as a book. None of this would have been possible without an Arts and Humanities Research Council (AHRC) Award, which gave me a unique opportunity to dedicate three years to pure scholarship. Additional travel funding from the AHRC, as well as fellowships held at the Kluge Centre at the Library of Congress and the Huntington Library, enabled me to complete in-depth research and a first draft of the manuscript. Emily Williams, Andrea Vesentini, and Lizzie Johnson patiently listened to me talk about energy over countless lunch breaks and kept me sane during these years. This period is best defined for me by Mark Cousins's infamous Friday Night Lectures at the Architectural Association, which demonstrated the power of ideas (and swearing) to move a room and upset the status quo.

Over the ten years since this book was conceived, many mentors and institutions have supported my career development and kept me in employment, which, as a result, has given me the necessary time to complete the project. When first invited to work on the AHRC collaborative project "Material Cultures of Energy" by Frank Trentmann, based at Birkbeck College, I thought there had been a mistake and that my skills in analyzing the language of energy left me unsuited to become a scholar working on carbon. However, this invitation (misguided or not) proved to be a gift, transforming my understanding of energy's discursive role in society. I owe a huge debt to Frank Trentmann for this baptism of fire, and his generosity in opening so many professional doors and passing on opportunities to me cannot be overstated. I am also grateful for the experience of collaborative research and how this enriched my practice as a scholar. For this, I would like to thank Hiroki Shin, Vanessa Taylor, and Heather Chappells for being such supportive colleagues.

This project also connected me with the emerging field of energy history, and in the process, it allowed me to reconceptualize the broader significance of the book project. I would like to thank the Joint Centre for History and Economics at Cambridge and Harvard and Paul Warde for allowing me to present material from this book within the context of energy history. More recently, I am pleased that I could argue for a place for William James as an integral figure in energy history, with a selection from this book appearing in the forthcoming collection *Energy's History: Toward a Global Canon*, edited by Daniela Russ and Thomas Turnbull.

The peripatetic years spent moving between organizations also provided the breathing room this project needed to grow. I would like to thank everyone in the Sussex Humanities Lab at the University of Sussex, where I was based as a research fellow in Mass Observation studies in 2017; in particular, I want to thank Caroline Bassett, Tim Hitchcock, James Baker, Fiona Courage, and Amelia Wakeford for their support and for facilitating such a dynamic environment through which to test out new ideas. I am grateful to David Huyssen and Sanjoy Bhattacharya for supporting my Joint Wellcome Trust/Centre for Future Health Research Fellowship at the University of York in 2018, which allowed me to lay the foundation for my next research project (while completing the final book manuscript).

I owe a large debt to Northumbria University, where I have been lucky enough to work since 2018. I could not have hoped for a more supportive and comradely institution to call my academic home. My thanks go to Daniel Laqua, Brian Ward, Matthew Kelly, and Charlotte Alston for their mentorship and support. I am fortunate to have a local community of Americanists and environmental historians to run ideas past and who have helped me better understand how to bring these two disparate fields together. Sitting in both camps, Elsa Devienne is owed special thanks for her astute feedback on the manuscript and for keeping this book on track through her rather fierce implementation of arbitrary deadlines. Hilary Francis, Jennifer Aston, Katarzyna Kosior, and Felicia Gottmann have lifted my spirits over many plates of dumplings and were there to tell me when the time was right to press submit. Northumbria University has provided extensive support for this project, providing additional funding and allowing me the time and space to complete it. I would also like to thank Matthew McAdam, Adriahna Conway, Kyle Kretzer, and Jeremy Horsefield at Johns Hopkins University Press, as well as the anonymous peer reviewer, for their editorial suggestions and advice. They have made the manuscript stronger. Dominic Boyer and Imre Szeman also need to be thanked for establishing the Energy Humanities book series and welcoming this book into it. I am very grateful for all they are doing to support and shape the development of this rapidly growing field.

The one constant during the writing of this book has been my friends and family. Cecily Machin, Caitlin Condell, and Vanessa Nicholas have let me sleep on couches and provided important distraction over dinners and drinks. My father has looked on bemused throughout this whole process and politely refrained from asking when the book was going to be out. I lean on my big sister, Louisa Wright, for everything, and this book has been no exception. Thanks to her and Chris Walker for always being there and to my nieces, Rosa and Asha, for making me laugh. Mark Byers has been on this journey since we were both research fellows at the Library of Congress together in 2013, where we met over a chocolate blondie during a research seminar where part of this work was presented. It is hard to believe that we have grown a family alongside this book and that we still have the energy to talk about energy after all this time. Both born in the final stages of this project, my children, Louis and Alice, have done nothing but delay its final publication. I wouldn't have it any other way. Finally, this book is dedicated to my mother, Karen Wright, whose stroke in 2018 has left her unable to read it. She remains, nonetheless, my perennial cheerleader and my biggest inspiration, and I know of no stronger woman.

Moral Energy in America

Introduction

> When the dynamite charge blows up human beings instead of
> rocks, when its outcome is waste instead of production, destruction
> instead of construction, we call it not energy or power but violence.
> —John Dewey, "Force and Coercion," 1916[1]

In the spring of 1916, as the American public debated intervention in World War I, one of the nation's leading intellectuals, the pragmatist John Dewey, came under attack from his former student and acolyte, the writer Randolph Bourne. The dispute resulted in one of the most embittered intellectual confrontations of the period, better known as the Bourne-Dewey debate.[2] The argument centered on whether the United States should enter the war in Europe, with the two figures taking opposing views. Dewey, writing in the progressive mouthpiece the *New Republic*, rallied support for American intervention in the war, seeing it as an instrument to expand democratic reform internationally. In a counterattack published later in the avant-garde journal *Seven Arts*, Bourne, who was against any participation in the war, tore apart this position, accusing Dewey and the entire pragmatist tradition of moral bankruptcy, of eliminating higher values in the name of efficiency.[3]

Given the high stakes of this debate, what is surprising is that, on closer inspection, this argument reads less like a battle of clashing ideologies and more like a semantic disagreement over a long-contested term—*energy*. Indeed, in the series of articles authored by Dewey, his argument depended on separating the meaning of energy from a rival term, *force*. On both sides of the debate, clarifying the meaning of energy provided valuable support for or against intervention in the war in Europe. For Dewey, the capacity to transform force into energy, to organize it to greater efficiency, provided the pragmatist argument for American intervention in the conflict. Bourne, however, pointed to Dewey's hypocrisy:

in mocking the isolationists' rejection of force, Dewey had confused energy with intrinsic value.[4] In the absence of an adequate body of ideals to set the energizing process in motion, Bourne claimed that Dewey, as well as the entire pragmatic tradition, had fallen for the fallacy that energy would de facto work for democracy.[5] Instead, Bourne insisted that because energy was not an ethical principle in itself society needed a separate field of values to channel it in the right direction. "This careful adaptation of means to desired ends," he argued, "this experimental working out of control over brute forces and dead matter in the interests of communal life, depends on a store of rationality, and is effective only where there is strong desire for progress."[6]

The moral justification for or against war depended, in this case, on clarifying two extremes on either side of the debate: one that took energy to be a universal value and another that understood energy to be plural, to be relational, valued against an end. Both positions raised the following question: Was energy external to social form, or did it only gain value as it was put to work in society? This was not just a technicality. Central questions about American democracy depended on its clarification. In Dewey's case, the potential to organize force into energy was reason enough to justify American intervention in Europe. For Bourne, value always preceded energy, and as a result, no magnitude of energy could support such an immoral cause as war.

At this moment of heightened tension, why Dewey and Bourne felt the need to spill ink refining the meaning of energy is telling. Far from being a simple matter of semantics, by the time the Bourne-Dewey debate erupted, energy had become a moral problem. This moralizing of energy was not just born of the recent crisis but crystalized debates that extended over a forty-year period, straddling the rise of progressive politics, the Great Depression and the New Deal, and two world wars. During this period of rapid social change, energy emerged in American thought as a central ethical problem. Energy had become moral.

The Roots of America's Energy Consciousness

When President Donald Trump pulled out of the Paris Climate Agreement on June 1, 2017, he validated a widely held belief that America lacks an energy consciousness. As the world committed to reducing carbon emissions, Trump was boasting about discovering a "newfound American energy" that "will power our ships, our planes and our cities."[7] Having long been a world leader in energy consumption (only to have been overtaken by China in 2010), the stereotype of the United States as a nation of oil-guzzling cars, overheated houses, and waste is more familiar today than it being a nation with a well-developed energy consciousness. Bob Johnson has even gone so far as to argue that Americans have largely "sublimated" their dependence on fossil fuels.[8] Even though fossil fuels have saturated every area of American life, public discourse about energy, Johnson points out, has remained remarkably scarce.[9] It was only when technological infrastructures broke down (or fuel prices rose) that fossil fuels became a visible presence to ordinary Americans. This explains why the oil crisis of the 1970s has often been identified as the origins of an American "energy consciousness," a moment when the nation could no longer ignore its dependence on fossil fuels.[10]

The assumption that America lacks an energy consciousness, however, overlooks a tradition of moralizing, conceptualizing, and politicizing energy in American thought. As the Bourne-Dewey debate shows, when one expands the definition of energy beyond a narrow focus on fossil fuels, one sees that energy has long been at the heart of American intellectual discourse. Throughout the Progressive Era and within the reform tradition that informed the following decades, energy was bestowed with a uniquely ethical character. Presidents, social reformers, scientists, economists, and public intellectuals all debated the nature of energy and its relationship to society. During this period, the problem of energy (its properties, how it might be harnessed, and how to qualify its social value) became central to a larger question about the condition of American society. Running through this conversation was a sense that energy in modern society was under threat—that it was running out, clogged, misdirected, or driven into the wrong channels. This anxiety emerged as a physical threat as fuel shortages, rate hikes, the misassignment of resources, and waste pointed toward a material blockage. But it was also expressed metaphorically, as society's energy appeared obstructed by a maladapted social, political, and cultural environment. In a forty-year period that shook American democracy to its core—a period that suffered through the Great Depression, the rise of communism and fascism abroad, two world wars, and the atomic bomb—attention turned to the nation's moral energy.

This energy language did not disappear with the detonation of the atomic bomb. Rather, it evolved over the post-1945 period and the decades that followed and continues to inform how Americans think about energy today. President Trump drew on this energy language in 2017 while plotting what he called the "golden era of American energy."[11] His promise that, in order to meet "the future of America's energy needs, we will find it, we will dream it, and we will build it" could have come from the pen of the American philosopher William James or a number of American presidents, from Theodore Roosevelt to John F. Kennedy, who each expressed concern about the vigor and energy of the American nation. Other presidents have instead focused on the control and direction of the nation's energy. Cold War concerns about America's lack of purpose, for example, led to the establishment of a President's Commission on National Goals in 1960 by President Dwight D. Eisenhower to provide "a sense of direction" for the nation's energy.[12] In a similar manner, President Jimmy Carter secured his electoral unpopularity by repeatedly urging the nation to "conserve its energy" for both material and moral reasons.[13] More recently, with William James's "moral equivalent of war" being used to describe the Green New Deal, we can see how this progressive political platform was grounded in an early twentieth-century discourse centered around redirecting or rechanneling the nation's energy. A distinctive energy language circulates across the political spectrum, underscoring a variety of programs from Trump's extractive economics to green politics. This book will chart how a particularly American way of thinking about energy evolved over the twentieth century. It will show that the United States has, in fact, a well-developed tradition of thinking about its energy.

In uncovering this discourse, this book will highlight the importance of conceptual frameworks in defining society's relationship to energy. To date, energy history has been largely disconnected from intellectual history.[14] The growing field of the energy humanities, however, has shown how carbon systems are embedded within complex cultural-symbolic and epistemological regimes. Important work by Cara New Daggett, Imre Szeman, Dominic Boyer, Bob Johnson, Stephanie LeMenager, Matthew T. Huber, and Timothy Mitchell, among others, has demonstrated that fossil fuels have driven the narratives, ontologies, subjectivities, epistemologies, political economies, and cultural forms of modernity.[15] LeMenager, Johnson, Brian Black, and Frederick Buell, for example, have illustrated the myriad ways in which oil has reconfigured American culture and society over the past century.[16] Others, such as Mitchell, Huber, and Caleb Wellum, have traced how oil shaped the political regimes and ideologies of late capitalism and neoliberalism.[17] As Huber has shown, the neoliberal hegemony that emerged in the United States in the 1980s was a direct result of the materiality of oil, which drove the "political structures of feeling" central to the "American way of life."[18] Oil was not the only technical regime to have driven major social and cultural shifts. Sam Halliday, David Nye, Jennifer L. Lieberman, and Carolyn de la Peña have captured how electricity provided, in Lieberman's words, "a fascinating conceptual tool that inflected how many Americans described and understood their world."[19]

These studies have centered carbon, or technical regimes, as the driver of cultural and intellectual production. From this perspective, intellectual frameworks are rooted in the material world and the technological systems that drove modernity. It is important to remember, however, that for most of the twentieth century the term *energy* remained largely disconnected from energy systems. Fossil fuels were referred to directly by their material typology—as bituminous or anthracite coal, kerosene, natural gas, or petroleum—or more generically as fuel and power. It was not until the 1970s that energy commonly became a shorthand to discuss the material basis of an energy system.[20] More often than not, when energy surfaced in public discourse, it had little direct referential connection to the energy regimes that transformed American life during this period. Abstracted from these material systems, energy took on a range of other cultural meanings.

One cannot, of course, separate this energy language from the energy revolution that occurred during these decades. As Stephen Kern famously catalogued, this period was subject to its own "energy crisis," "albeit a crisis of abundance" brought about by new energy technologies that fundamentally "revolutionized the experiences of time and space."[21] The impact of new technologies, transport networks, and electricity systems drove the cultural production of the period, best encapsulated by modernism and its many revolutions of form. Moreover, as the field of energy humanities has taught us, in a world in which fossil fuels operate as, in the words of Szeman and Boyer, a "'spontaneous consent' of hegemony" on our lives, it would be a fool's errand to extract the imprint of different technical regimes from the myriad linguistic and epistemological models that surround us.[22] While we might not be able to isolate the language of energy (or indeed language itself) from its material underpinnings, we can identify an alternative energy language that emerged out of discourses far removed from the fossil economy. It is only once we turn our attention away from the explicit reference to energy's material regimes that we can begin to identify the roots of a moral discourse around energy that also shaped the American imaginary.

A growing body of scholarship has documented how societies have thought about energy beyond its materiality. Scholars such as Anson Rabinbach, Barri J. Gold, Crosbie Smith, and Allen MacDuffie have mapped the transport between scientific and cultural understandings of energy, examining the circulation of thermodynamic allegories within cultures of modernity.[23] Cara New Daggett has gone farthest to outline a genealogy of energy. In doing so, Daggett has demonstrated that energy operates as a master narrative, "a set of tropes and metaphors that help to describe a 'historical kind of interaction,' one that is continually generated at the intersection of bodies, machines, and fuels."[24] Daggett provides a compelling account of how a distinctive energy language emerged out of Victorian Britain and came to shape energy thinking in the West. The energy language that Daggett unearths, however, remains tightly bound to nineteenth-century science and the laws of thermodynamics.[25] In doing so, Daggett provides an important master narrative that explains how energy drove a new ethical paradigm centered around work that continues to inform our energy epistemologies today. Daggett is not alone in locating energy at the origins of this paradigm shift. Rabinbach's account of the emergence of the "human motor" positions the science of thermodynamics at the root of modernity and our understandings of labor and the body.[26] These well-defined accounts bind energy to work and to the rise of a dominant social order organized around capital, efficiency, and environmental exploitation.

Although both the technological and the thermodynamic framings allow us to understand how, and why, energy became a dominant paradigm in the West, they also risk concealing the other contexts in which energy epistemologies have been set and the multiple meanings that came to be attached to them. As this book will show, energy was not restricted to technological regimes or the laws of thermodynamics. And its meanings extended far beyond that of work. Instead, energy was absorbed into broader intellectual debates that circulated through America in the first half of the twentieth century. In the case of the Bourne-Dewey debate, energy was situated at the heart of a wider conversation over the merits of pragmatism and idealism, as well as issues surrounding pacifism, internationalism, and democratic reform. Energy also surfaced within broader social debates around free will, social reform, the climate, race, urban planning, economic value, ethics, and the changing landscape of radical politics. Moreover, energy was situated within a tradition of American thought that extended back to Jefferson and forward to midcentury ideas about Technocracy and the world state. By widening the scope to include discourses only apparently tangential to energy's thermodynamic and material regimes, this book will uncover a body of energy discourses that emerged out of a distinct American intellectual tradition that was closely linked to contemporary social concerns.

Uncovering the intellectual contexts in which energy took root is important not just for understanding how energy as a concept circulated; it also destabilizes many of the tropes that have come to shape the ways in which America's relationship to energy has been narrated. This has often been equated to "a love affair," albeit a highly addictive and destructive one. Its meteoric rise as one of the world's leading energy consumers has meant that there is an internal momentum to narratives of America's energy dependency. Once it broke from the organic economy, the story goes, the United States was predestined to emerge as the high-energy society it is today. Tropes such as "the great acceleration" give the impression of a system growing beyond human control and determination.[27] In these accounts there appears to be little resistance to the dominance of energy across every domain of American life. Americans encountered these new technologies with open arms—ready, as Bob Johnson states, to "jump into modernity's orgy of production and reproduction without regard to organic limits."[28] These accounts grant a unique agency to fossil fuels as they swept through American culture, remaking it in their image. Historians have picked apart this deterministic narrative to reveal the complex social factors that shaped the rise of America's energy dependency.[29] The growth of America's energy systems, it has been shown, was dependent on a range of local, environmental, economic, political, social, and cultural factors. Less attention, however, has been paid to the intellectual landscape within which this energy revolution took place.[30]

As this book will show, these energy systems did not emerge from an intellectual vacuum. Instead, a public discourse on energy grew out of an American intellectual tradition focused on the state of American society and its institutions. As we shall see, people struggled to come to terms with how to reconcile energy with preexisting intellectual debates, ethical systems, and political ideas. This public discourse on energy, moreover, remained deeply concerned about the moral value of energy to American society. Time and again thinkers tried to pin down a connection between energy and social good and, as a result, constructed elaborate theories about energy's relationship to society. Even though this conversation was often not directly referring to material energy regimes, it became a model through which people understood and conceptualized their connection to this abstract thing called energy.

Through this discourse we can also begin to identify precursors to current critiques of energy determinism and calls to decouple energy from growth.[31] The connection between energy and work has a long intellectual legacy, as Daggett has traced, which extends from early debates surrounding the science of thermodynamics to being embedded within current economic models, such as GDP.[32] This book will unearth a counterdiscourse that has long challenged this false equivalence. Buried within the intellectual debates of the early twentieth century was a powerful critique of energy's intrinsic value. Rather than a blind celebration of energy, this discourse placed the moral onus on society to direct, control, and organize its available energy. This led to an alternative energy ethics that stressed the importance of ends, direction, and the harnessing of energy for social good. Of course, what constituted social good (or higher ends) remained relative and often involved the exclusion and management of certain social groups, such as women, immigrants, and African Americans. Through this ongoing debate, however, we can start to identify a counterdiscourse that remained alert to, challenged, and posed alternatives to the powerful energy metaphors that shaped the twentieth century. Locating a more nuanced public discourse on energy, therefore, will complicate the tropes surrounding America's relationship to energy and make way for new paradigms to define it going forward.

This is not only important for shaping how we understand America's relationship to energy. Building up a fuller picture of the intellectual cultures in which the fossil economy emerged destabilizes many of the narrative features at the heart of the Anthropocene. There is little doubt about the disproportionate impact that the United States has had on the growth of the Anthropocene, a term first coined in 2000 by the biologist Eugene Stormer and atmospheric chemist Paul Crutzen to describe a new geological epoch driven by mankind.[33] The statistics say it all. By 1965, the United States consumed one-third of the world's energy.[34] And yet, even though the material impact of the United States on the Anthropocene is not in doubt, the narratives that surround its development demand closer inspection. Andreas Malm has been a prominent critic of what he terms the Anthropocene narrative.[35] This is the narrative that describes the Anthropocene as an inevitable outcome of human mastery over nature and the innate need for human progress and growth.[36] Critics of the Anthropocene narrative highlight the totalizing nature of this account, removing the dynamics of culture and power from the story. By decontextualizing the rise of fossil fuels, Malm points out, they *"renaturalised"* a historical process that was bound up in broader social contexts.[37] Recent postcolonial critiques of the Anthropocene narrative have demonstrated how divergent fossil fuel pathways occurred outside of the West. Focusing on the Global South and Asia, historians such as Elizabeth Chatterjee, Victor Seow, and Antoine Acker have plotted new stories that map the growth of fossil economies, stressing the importance of microhistories so as, in Acker's words, "to imagine an Anthropocene with several departure points."[38] Not only, however, do we need to geographically decenter the dominant Anthropocene narrative, but we also need to build up a fuller picture of the intellectual contexts in which fossil fuel regimes were adopted in the nations at the forefront of the Anthropocene's development.

This book will show that public discourse around energy complicates the Anthropocene narrative. As we shall see, there existed an extensive debate about energy's moral value. This public debate was critical in framing how Americans understood their relationship to energy and its role in society. There was nothing innately "Anthro" about the American experience, but it was rooted in an American way of thinking about energy grounded in a specific intellectual tradition and set of concerns that dominated the first half of the twentieth century. Energy, as this book will demonstrate, was woven into broader public discourses around the nature of the American state, its racial constitution, its ethical systems, its economic and cultural foundations, and its political forms. By exploring how energy was set within these wider discourses, we can begin to see how deeply our fossil economies, including those at the forefront of the Anthropocene's development, are rooted in localized intellectual cultures and traditions that also need to be attended to.

This book, finally, argues that environmental history, energy history, and even the energy humanities would benefit from paying closer attention to the history of ideas. As Paul Sutter has argued, in the past few decades environmental historians have "retreated from engagement with intellectual history in favour of a 'cultural turn.'"[39] Despite first-generation environmental historians such as William Cronon having a close affinity with the history of ideas, over the decades, Sutter argues, environmental historians have become "suspicious of the realm of ideas in general" and have increasingly disconnected their work from intellectual history.[40] Of course, as John Toews famously pointed out, after the linguistic turn "we have all become intellectual historians" to some degree, critically engaging with language and the production of meaning and its impact on the material world.[41] Despite this, environmental and energy historians have tended to overlook "the history of ideas" as a discrete site for understanding how our relationship with the environment and fossil fuels has been established. Even the field of energy humanities, for example, which has pushed the importance of energy epistemologies to the fore, has yet to fully uncover energy's place within the history of ideas (beyond its scientific or technical definition).[42] Christopher F. Jones has diagnosed what he calls a "petromyopia" within the energy humanities, whereby the focus on petroleum has overlooked the dynamic interplay of other energy sources.[43] One could extend this further to suggest that the focus on energy as a technological or scientific entity has also covered up the broader intellectual landscape in which ideas about energy have been set. As a result, intellectual history continues to be overlooked as a site for unearthing our relationship to energy.[44] This book will demonstrate that broader intellectual currents had a definitive role in shaping how Americans understood and conceptualized their relationship to energy. As such, it will argue that more attention needs to be given to ideas when examining the roots of our current energy and environmental crisis.

Debating Energy's "Moral Value"

At the heart of this public discourse about energy was an ongoing debate about its moral value. As many commentators recognized, energy as an idea was slippery, especially when trying to determine its value. This book will identify two distinct positions that would come to define how energy was attributed moral value: one that viewed energy as an objective measure for society and another that put the emphasis on society to transform energy into social good.

To grasp how these positions developed, we first need to understand a fundamental contradiction that emerges when one views energy as moral. That is, to describe energy as moral undermined energy's most unique quality—that energy escaped moral value. Following the emergence of the science of thermodynamics in the mid-nineteenth century, energy assumed this allure, as it was understood as a direct register of nature. To the founders of the conservation laws, James Prescott Joule, Julius Mayer, and Hermann von Helmholtz, energy placed economics at the heart of nature, making thermodynamics a science for accountants eager to balance nature's books.[45] The conservation law reduced all of nature to an all-inclusive energy or *Kraft*. Once this was settled, it was not long before all materializations of force—in nature, the machine, the human body, and society—could be measured as part of a universal force, an objective record of nature.[46]

Behind the assumption that energy was an objective record of nature rested the supposition that a magnitude of energy reflected a constant economic value. This depended on a worldview that took nature to be industrious—an outlook Anson Rabinbach termed "modern *productivism*."[47] Embodied by the metaphor of the "human motor," this prospect depended, Rabinbach explains, on a "belief that human society and nature were linked by the primacy and identity of all productive activity, whether of laborers, of machines, or of natural forces."[48] Within this productivist worldview, energy was taken as a direct correlate of value, as the fecund and creative energy of nature worked in a universal direction, driving modern industrial society as smoothly as it would a steam engine. By reducing nature to a universal value, energy provided the appearance, in the words of Philip Mirowski, that "a bookkeeper had finally seen beyond the veil of money into the real world of fundamental values, no longer having to worry about inflation, arbitrary depreciation, and the like."[49]

Energy clung to its privileged position as an objective standard of nature. The engineering and managerial ethos that emerged during the Progressive Era and lasted until the Second World War drove a generation of reform-minded social scientists, administrators, intellectuals, journalists, artists, and writers to turn toward applied science as the basis for a new social and political culture.[50] During this period, a common creed emerged that a healthier society could be created through greater organization, efficiency, and rationalization.[51] People looked self-consciously to the machines that pervaded everyday life—recognizing in these technologies the potential for a better society grounded in a "machine-age consciousness."[52] During this period, as John M. Jordan has demonstrated, the machine became a template for a society run by cause and effect, built on science rather than subjective whim or politicking.[53] This managed society, which grew under the shadow of Ford's assembly line, strove to eliminate all "human elements" that failed to accord with the logic of equivalence put forward by the energy laws. Not only would the time-motion studies central to Frederick Winslow Taylor's scientific management reduce the human to a carefully regulated thermodynamic machine, but private interests, sentimentality, religion, and idealism would be discarded along with supposedly volatile and intransigent social groups, including women, immigrants, and African Americans.[54] In this context, energy came to be privileged against speculative forms of knowledge generation: illusory market values, unreliable politicians and businessmen, and metaphysical or theological dogma.

As the social sciences and reform politics turned to the shared lexicon of objectivity, rationality, and equivalency, energy became a cherished social measure, a "yardstick" immune to social corruption and manipulation.[55] The Technocracy movement swept the national stage in 1932, promising an economic system backed by energy rather than money. This guarantee was preserved in President Franklin D. Roosevelt's political rhetoric, as energy came to embody the stability guaranteed by the New Deal. For over two decades, energy was synonymous with a range of positive attributes that celebrated it in relation to a set of negative values. Commentators championed its best features, which included its claim to the natural, the intrinsic, the objective, the a priori, and the factual. Blessed with these qualities, energy continued to be cherished as the gold standard of nature.

Yet just as energy was celebrated for its objectivity, for its privileged position outside of social forms and institutions, the paradox remained: if energy was to have any value, it had to be harnessed by society into work. In other words, it had to be converted into social value—into a moral equivalent. Nowhere had this become more apparent than when the United States confronted the conflict in Europe. In 1915, the *New York Times* published Henri Bergson's forecast about the future prospects of Europe's "moral force."[56] In his usual prophetic style, Bergson characterized Germany as a cult of brute force, of rampant materialism that had replaced moral force with "the mechanization of spirit."[57] The privileging of material over moral force, Bergson believed, would lead to Germany's defeat, as its worship of "might makes right" left the nation with only material energy for backing, living on reserves rather than self-replenishing ideals.

The "cult of force" surrounding Germany, therefore, became a caricature of the brute worship of energy. Dewey did much to foster this stereotype, arguing that although the "gospel of the strenuous life," "of the value of energy or will for its own sake," had been taken as a distinctly American attribute, this was unique to the German sensibility.[58] Remaining true to the means-ends rationale, the American, Dewey explained, required an end for his energy, "from winning a ball game, or forming the biggest business corporation in the world, to converting a community to Billy Sundayism."[59] Entrenched in the language of Romanticism, Germany celebrated "an absolute energy striving through personal channels for manifestation." When this was transferred from the individual to the state, it supplied "the current Teutonic apologia" for the present war.[60]

The spectacle of war thus worked to undermine any objective value attributed to energy. Germany's aggressive materialism and its Romanticism had illustrated to the American public the dangers of worshipping energy for its own sake. Furthermore, not only were rival nations busy willing energy to work for them, but there was also little consensus as to what, if any, natural value energy had. Following the war, Sigmund Freud complicated this problem, further imbuing energy with a Janus-like character. In "Beyond the Pleasure Principle" (1920) Freud assigned two contradictory drives to energy: one productive (Eros) and the other destructive (Thanatos, the "death drive"). By undermining any moral *priorism* attributed to energy, Freud created an indelible tension whereby "one group of instincts rushes forward so as to reach the final aim of life as swiftly as possible; but when a particular stage in the advance has been reached, the other group jerks back to a certain point to make a fresh start and so prolong the journey."[61] Although this strain was implicit to Freud's earlier libido theory, his model of the "death drive" formalized energy's moral ambivalence.[62] While the war had confirmed the existence of this destructive drive on a mass scale, growing fears over the rationality of the democratic subject lodged a negative energy at the heart of society.

Alongside energy's undetermined morality, another central confusion arose when observers tried to calculate energy's ethical value. Here the promise of equivalency broke down as people struggled to correlate a quantity of energy with its quality—its social value. The philosopher William James foregrounded this problem, cautioning that energy's social worth lay in its quality rather than its quantity.[63] In his 1906 American Philosophical Association presidential address "The Energies of Men," James had stressed the importance of "qualitative" over "quantitative" uses of energy, as he turned his attention toward the problem of "mental" and "moral" work.[64] Establishing a hierarchy of "energizing," James pointed out that "our muscular work is a voluminous physical quantity, but our ideas and volitions are minute forces of release, and by 'work' here we mean the substitution of higher *kinds* for lower *kinds* of detent. Higher and lower here are qualitative terms, not translatable immediately into quantities."[65] Because "higher and lower" were qualitative, unable to be translated into a magnitude, elevated ethical states like "having a more elastic moral tone" upset any natural value granted to energy by physics.[66] Thinking, James maintained, was of a higher social value than walking, "even though the total heat given out or the foot-pounds expended by the organism, may be less."[67] Because of this lack of continuity, James noted elsewhere, "the term 'energy' doesn't even pretend to stand for anything 'objective'! It is only a way of measuring the surface of phenomena so as to string their changes on a simple formula."[68]

Despite this misconception, energy remained a conceptual device central to James's pragmatic philosophy, and in his speech "The Moral Equivalent of War" he supplied a blueprint to the following generation of reformers as to how a "moral equivalent" could channel energy into elevated ethical forms.[69] He had already raised this point in *The Varieties of Religious Experience* (1902), where he argued, "One hears of the mechanical equivalent of heat. What we now need to discover in the social realm is the moral equivalent of war."[70] Through a "moral equivalent," energy could undergo a qualitative transformation, sublimated from negative pursuits like warfare toward more productive ends—in James's case, a war against nature.[71]

This call resounded through the following decades. In 1913, the progressive journalist Walter Lippmann advocated a novel form of "state-craft" to harness the nation's energy into "fine values."[72] Turning to James's "moral equivalent" and combining it with Freud's libido theory, Lippmann maintained that "the energies of the soul" are "neither good nor bad in themselves": "Like dynamite, they are capable of all sorts of uses, and it is the business of civilization, through the family and the school, religion, art, science, and all institutions, to transmute these energies into fine values."[73] In the following decades James's "moral equivalent" would be applied to all fields of social life, with gardening, farming, urban playgrounds, football, and a range of civic betterment agencies proposed as moral equivalents to transform society's wayward energies into higher forms.

Even though, then, energy would come to epitomize the objectivity of the machine age, it remained a fundamentally moral proposition. Rather than being a universal value found in nature, for energy to have any social worth, as many like James recognized, it had to be relational, it had to be put to work for society. In other words, far from being an objective entity outside of society, energy only gained value when it was put to work. As James himself put it, a hydroelectric ram might go "clap, clap, clap, day and night . . . raising so many kilogrammeters of water. What the *value* of this work as history may be, depends on the uses to which the water is put in the house which the ram serves."[74]

Two broadly defined positions, therefore, shaped the debate about energy's moral value. On the one side were those who looked to energy as an ethical standard against which to measure American society. From this perspective, energy would be privileged because of its objectivity and its capacity to escape subjective value. On the other side were those who maintained that energy was qualitive— that it only gained value through direction, through an end. From this perspective, the emphasis fell on American institutions and moral values to harness energy toward socially beneficial ends. One position looked to energy as a blueprint for American society, whereas the other stressed the importance of building a social and ideational infrastructure strong enough to manage and direct energy to constructive ends. Even though these positions would never be entirely distinct and would often overlap and become confused in application, they drove the debate about energy's moral value in the first half of the twentieth century.

Identifying Energy Discourses

What also united these positions was the fact that energy remained an abstraction. Indeed, energy came to be such a usable concept precisely because it was a metaphor, bearing little relation to its physical properties. Whether it was bolstering reformist politics, anti-immigrant sentiment, or laissez-faire economics, energy took on a privileged role precisely because it became nothing like energy. As it was harnessed into cultural work, energy lost all resemblance to the physical entity it was supposed to represent. This led to what I will term the *energy paradox*: the closer people strove to capture the qualities of energy, the more abstract energy became. Burdened with this contradiction, geographers set out to reduce incommensurable social forms to energy, only to resort to arbitrary equivalences; urban planners tried to distinguish primitive energy from industrial flows, only to realize that they were intertwined; and economists sought to reduce everyday products, like bacon and eggs, to an energy value, only to confront the fact that when it entered the market energy became little different from money. Alongside this, sociologists struggled with the fact that, as an absolute measure, energy forced the plurality and diversity of the political, economic, sexual, and spiritual life of the nation into an arbitrary common standard. How, they questioned, could one reduce the varied and intangible emotions embodied in conditions such as fear, lust, excitement, creativity, and despair into a quantifiable metric standard as restrictive as energy? For others, energy became a social leveler, a structural paradigm, a unitary agent, and even, in some cases, an authoritarian dictator. Energy spoke to progressives, conservatives, radicals, liberals, fascists, and free market advocates alike. Energy became anything people wanted it to be—except, of course, energy itself.

This book turns its attention to this range of thinkers who were drawn to energy for very different reasons. Mavericks, eccentrics, and aficionados sit beside larger social movements and political programs. Scientists, social reformers, intellectuals, politicians, economists, writers, geographers, and even radio enthusiasts were united through their mutual attraction to energy. There are important reasons for including such a broad range of figures in a study that looks to reconstitute an intellectual history of energy. During its long history, energy has problematized not only disciplinary boundaries but also central divisions within Western philosophy, such as the mind-body dualism. This was why energy was so appealing to Monists seeking a unified framework through which to understand the world.[75] For the German chemist and founder of energetics Wilhelm Ostwald, who headed the German Monist League from 1911 to 1915, energy could describe all aspects of reality: his catchphrase "waste no energy; turn it all to account!" provided a theory of happiness, of civilization, of the economy, and of all aspects of life.[76] Although Ostwald's energetics remained a short-lived project, the efficiency movement that rose to prominence during the Progressive Era came to dominate the American imagination. During this period, the science of energy swept through the human sciences and into people's homes, influencing a variety of activities, from cooking to managing leisure time, extending into moral and political behavior.[77]

Not only did the expansiveness of energy extend across all social domains, but it also spilled across geographic borders. Although this book focuses on the United States, it also captures the considerable transatlantic interaction during this period. Significant transatlantic contact occurred throughout the scientific community, albeit interrupted during the war years, with figures like the German-born biologist Jacques Loeb continuing an extensive correspondence from America with scientists across Europe. Reform ideas from Europe also filtered into the United States.[78] Lewis Mumford traveled to Edinburgh in order to meet (although he didn't, as it turned out) his mentor Patrick Geddes, just as the British urban planner Ebenezer Howard arrived in the spring of 1925 at the Hudson Guild Farm in Mount Olive to join the Regional Planning Association of America (RPAA) at their annual gathering.[79] Travel was not the only way ideas spread. The American press frequently reported the ideas of British intellectuals in newspapers, magazines, and periodicals and (after the mid-twenties) over the radio. The British scientist Julian Huxley, for instance, could often be found in the pages of the *New Republic*, the *Nation*, and *Harper's Monthly* discussing his theory of planned democracy from the mid-twenties all the way to his appointment as director general of the newly founded United Nations Educational, Scientific and Cultural Organization (UNESCO) in 1946.

Although energy metaphors traveled freely between communities and across borders, it is important to recognize that this energy language remained restricted to particular social groups. Those who approached the subject of energy did so largely from a white male perspective. This was partly because the figures responsible for cultivating a new discourse on energy, such as Theodore Roosevelt, Brooks Adams, and William James, were responding to the crisis of masculinity that saturated American culture at the turn of the century. Energy, therefore, surfaced as part of a broader gendered and racial discourse used to legitimize cultural superiority over non-whites and women. These associations clung to energy in the following decades, as they informed universalist conceptualizations of Western civilization and "modern man." Chapter 2, for example, will illustrate how energy paradigms operated as central models through which to highlight racial difference and concerns around immigration and racial degeneration during the interwar period. It is important to note, therefore, that this energy language did not belong to everyone and was used to reinforce hierarchies of power that circulated at the time.

This book follows the evolution of this moral discourse of energy over the first half of the twentieth century, from its origins in the Progressive Era to its saturation in the social imaginary of the interwar period and the New Deal and its redefinition in the years preceding and following the Second World War. Each chapter focuses on a distinctive area in which energy became prominent: from the political debates of the Progressive Era to interwar understandings of the climate and race. In doing so, each chapter identifies a distinctive energy discourse that illuminates the ways in which energy informed intellectual debates far removed from the laws of thermodynamics and the fossil economy.

The book starts by tracing the origins of a moral discourse of energy within the political thought of the Progressive Era. Chapter 1 will reveal how, in the years leading up to the First World War, a discourse around what I term "progressive energy" became part of a wider battle to define the soul of the nation. The chapter traces how energy became central to an ongoing debate about the role of the individual will in the industrial period and its repercussions for progressive political thought. Whether or not the will was subject to energy laws, the chapter will show, was not just a scientific debate; it raised fundamental questions about the state of American democracy, the individual, and the progress of the nation. To monists such as the physiochemist Jacques Loeb and the chemist Wilhelm Ostwald, the reduction of all phenomena to energy, including the will, provided a powerful model for a collectivist and socialist model for society. In opposition, intellectuals, such as the psychologist and philosopher William James, placed the onus on the will as a generator of energy and stressed the role of the individual in a democratic society. This energetic model would support Theodore Roosevelt's mantra of the "strenuous life," a paradigm that came to define the cultural politics of the Progressive Era. Out of this debate, this chapter will show, came two broadly defined ethical positions that would shape understandings of energy in the decades to come: one that looked to energy as an ethical standard for society and another that placed the onus on American institutions to harness energy into form.

While chapter 1 locates the roots of a moral discourse around energy in the Progressive Era, the next three chapters show how this matured during the interwar decades. Chapter 2 will trace how a discourse of "climatic energy" became central to a wider discussion around race and Americanness in the interwar period. Energy, the chapter will show, became incorporated into a broader social debate about immigration, eugenics, and the preservation of the white Anglo-Saxon body. The chapter will examine how climatic determinists, such as the popular geographer Ellsworth Huntington, drew on an energy language as a way of measuring humans against an external measure in the environment. This energy discourse had ramifications for the way in which the environment was conceptualized. By the 1930s, as scientists started to register a warming climate, it was human energy—and, as a result, Americanness—that was seen to be under threat. Early fears about climate change, therefore, reflected broader anxieties about social degeneration and racial purity, all framed through a specific understanding of climatic energy. How far energy was an external measure for American society, however, divided people once again. While determinists like Huntington used it to categorize different racial groups, reformers stressed the importance of direction and the need for a strong physical and moral infrastructure to control and discipline the vigorous energy provided by the North American climate.

The next two chapters situate energy against the backdrop of the crisis in capitalism brought about by the Wall Street Crash and the Great Depression. Chapter 3 will uncover how a discourse of "psychic energy" became central to a broader cultural debate about the revitalization of American culture in the interwar period. Energy metaphors, emerging out of modern psychology and the work of Sigmund Freud, spread far beyond the psychological establishment to influence a group of planners and cultural theorists, including those closely affiliated with the RPAA. Prominent members of this group, which included the social critic Lewis Mumford and the polymath Benton MacKaye, were drawn to psychic energy as a way to understand the physical environment, informing the way in which large-scale infrastructure projects, such as the Tennessee Valley Authority, were envisaged and described. And yet, as planners, such as Mumford, conflated the physical environment with the landscape of the psyche, they confronted the energy paradox: just as they sought to locate a "primitive" energy as a blueprint for American society, they recognized that it was impossible to isolate an a priori energy from the environment.

Chapter 4 will examine how energy became subsumed within a larger discourse about economic value during the New Deal period. This was initiated by the Technocracy movement, which rose to fame in 1932 calling for a new energy unit of value for the monetary economy. This discourse was not restricted to those who sought to eradicate money, however. Franklin D. Roosevelt also drew on energy's promise as a social leveler, along with movements across the social spectrum, from monetary reformers associated with Social Credit to fascist sympathizers such as Ezra Pound. Even anti–New Deal advocates of the free market incorporated allusions to energy within their reform-based programs. As many critics recognized, this required a stubborn blindness to the fact that (as soon as it was proposed as a unit of exchange) energy was returned to the sphere of value, in some cases even transformed into its own ersatz currency. By the end of the decade, as social planning became associated with totalitarian regimes in Europe, free market liberals like Isabel Paterson conflated energy and money and thereby reformulated energy within an objectivist ethics for the market.

Chapter 5 will conclude by looking at how the changing political landscape leading up to and immediately following the Second World War created a new discourse on "energetic ethics." As the growing neurosis over fascism, totalitarianism, and atomic fallout came to characterize the period, the chapter will show how the flexibility of energy was extended toward the ultimate planned state, unifying the globe, while supporting an atomized politics that placed the individual at the center of an energy system. With concern fueled by the Frankfurt School about new forms of authoritarianism emerging in the United States, figures such as Lewis Mumford, who had once been one of the most vocal advocates of reworking society's energy, looked inward, focusing instead on the "rebirth of man." The heroic individual able to control and direct his energy became the celebrated cultural form of the period, taking on a greater prominence in Mumford's writing, as well as in the work of leading figures of the avant-garde, such as the abstract expressionist painter Jackson Pollock, the poet Charles Olson, and the writer Aldous Huxley. And yet, just as this energy language turned inward, a parallel discourse emerged calling for the development of a supranational organization, a world state, to channel the world's energy. In fact, the British scientist Julian Huxley, who also fueled the debate about the "rebirth of man" in the United States, pioneered a distinctive energy ethics that he would apply to one of the leading postwar international organizations, UNESCO. These competing discourses, one centered on rebuilding the individual and the other centered on refashioning the political institutions of the postwar world, placed the onus on "ethics" to manage the energy of the post-1945 period.

By identifying five different energy discourses, I hope to demonstrate how closely embedded energy was in the intellectual currents that defined American thought in the first half of the twentieth century. I will show that energy sat at the center of a broader public debate about the state of the United States, at a time when American progress, freedom, and democracy were perceived to be under threat. In doing so, I will argue that we cannot understand the contexts in which energy systems are set without recognizing the broader intellectual discourses that shaped their development. Only by exposing energy's intellectual foundations can we start to grasp the meanings we apply to energy and open up new pathways to reimagine them in the era of global warming and environmental collapse.

1908–1919

CHAPTER ONE

Energizing the Will

Progressive Energies

In 1909, America's leading physiologist Jacques Loeb wrote to the then president Theodore Roosevelt to advise him about the significance of the laws of thermodynamics for modern governance.[1] A German-born Jew, Loeb had relocated to the United States in 1891 to become the leading representative of positivist science.[2] He was a national celebrity, famously reported to have created life in a test tube after he invented artificial parthenogenesis by fertilizing a sea urchin egg with salt solutions. He was a pacifist, was politically aligned with the Left, and, although not adhering to any specific political movement or party, advocated a social democratic organization of society.[3] In his letter to Roosevelt he commended Roosevelt's administration for its adherence to "fundamental laws in pure science."[4] It was a shame, however, he added, that the "conservation of energy" had been discovered "only eighty years prior."[5] "Had the statesmen realized the bearing of this law upon the economic welfare of the nation," he continued, "they would have saved the sources of energy of the sun stored up in water power, coal and oil fields, forests, etc. to the whole nation, instead of allowing individuals to appropriate them, and thus partially enslaving all coming generations."[6] This would have resonated with Roosevelt, who was in the final months of his presidency. His administration was the first government to implement a widespread conservation agenda, which included an extensive program centered on the federal management of natural resources.[7] By offering Roosevelt a loaded compliment, Loeb used energy to prove a political point. It would be science controlled by energy laws, rather than "the limited horizon of the business man," that would dominate modern statesmanship.[8]

Loeb's admiration was short-lived. Not long after his laudatory letter he turned his attention to a more dangerous form of energy being promoted by Roosevelt.

"Power of Will" Surmounts All Obstacles

Advertisement for the book *Power of Will* by Frank Channing Haddock. "'Power of Will' Surmounts All Obstacles," *Popular Mechanics*, June 1916, 11.

This was best epitomized by the mantra of the "strenuous life," an ethos Roosevelt had popularized ever since he led the Rough Riders into battle during the 1898 Spanish-American War.[9] Playing the role of warrior, explorer, and huntsman to a captivated nation, Roosevelt had become the standard-bearer for a new archetype of modern masculinity, dependent on iron determination and a capacity of the will to overcome obstacles.[10] Described as "an electric battery of inexhaustible energy" and "the most vital man on the continent, if not the planet," Roosevelt feted an energy that flew in the face of hard science, drawing on a form of vitalist energy emergent from the will.[11] This mantra had taken on a new prominence in American life, as Roosevelt whipped up support for American intervention in the war in Europe. Far from enamored with Roosevelt's "electric" will, Loeb was adamant that this renewed emphasis on the will was fueling a wider social malaise—driving corrosive forms of nationalism, militarism, and capitalism.[12] This led Loeb to worry that it would be the man with the "perpetual motion machine" rather than the one schooled in "Helmholtzian physics" who would take hold of American culture.[13]

The fact that one of the nation's leading scientists felt the need to advise the sitting president on energy laws demonstrates how extensively energy had become part of a larger battle to define the soul of the nation. Loeb was not alone in recognizing this. Rather, energy had become central to a broader conversation about what ethical, social, and political qualities America should adopt as it confronted the war in Europe. The long-running controversy between mechanism and vitalism provided the foundation for this debate, foregrounding the question of the will and its relationship to energy. Whether or not the will could be reduced to energy laws not only was a scientific question but also fed into wider debates about American democracy, the individual, and the progress of the nation, all issues hotly contested during these politically tumultuous decades.

This chapter will trace the emergence of a moral discourse around energy during the Progressive Era, which I will call *progressive energy*. Energy saturated American culture during these decades. This was hardly surprising. Over the course of one lifetime, from the mid-nineteenth century to the turn of the century, new energy technologies had entirely reordered American life, allowing the nation, in Bob Johnson's words, to "sidestep the logic of the organic economy" through the release of a vast amount of mechanical power.[14] During this period, America's total energy consumption doubled every ten years, bringing with it new transport networks connecting the country, mechanized agriculture and industry that rapidly expanded the economy, and the transformation of the nation from a largely agrarian society into an increasingly urbanized one.[15] Electricity also became part of the public sphere. By the turn of the century Americans were able to enjoy a trip on an electric streetcar to visit a movie theater or an amusement park, or simply to view the electric lights that had become a feature of American cities. As a result of these dramatic material changes, American culture was rife with allusions to energy, from electrical metaphors to thermodynamic allegories. In particular, the second law of thermodynamics, which predicted the "heat death" of the universe, captured the American imagination, undermining a sense of unlimited growth and predicting a wider national decline.

The debate that emerged around progressive energy was a product of this cultural obsession with energy, driving a wider ethical debate about the place of the will in the industrial age. This chapter, therefore, will build on the work of Carolyn de la Peña, David Nye, Stephen Kern, and Tom Lutz, who have each shown how important energy was to the American imaginary during these decades.[16] It will demonstrate, however, that the fixation with energy had a distinctly moral character, feeding into a wider cultural debate about the place of the individual in American society and what this meant for the development of the political and ethical frameworks of the nation.

The chapter will start by tracing how this debate found its most extreme exponents in the work of the monists, such as Jacques Loeb and the physiochemist Wilhelm Ostwald, who subsumed the will under the energy laws to promote a social democratic and collectivist politics for the nation. It will then go on to explore how, in the hands of the nation's leading thinkers, such as the paleontologist Henry Fairfield Osborn, the political scientist Brooks Adams, and the historian Henry Adams, negotiating the relationship between energy and the will became central to competing social visions for the nation, from promoting social Darwinism to expressing wider anxieties about degeneration and national decline. Finally, it will turn its attention to the intellectual giant William James, who did the most to foreground the importance of the will in harnessing energy. James placed the onus on the will to direct the energy in the universe. In doing so, he promoted a new ethics that placed the individual at the heart of the energy system, a position most famously popularized through Theodore Roosevelt's promotion of the "strenuous life."

The debate about energy's relationship to the will, moreover, laid the foundations for two opposing positions that would inform debates about energy in the decades to come: one that turned to energy as an objective ethical standard able to eliminate subjective categories (like the will) from the social order and another that placed the moral onus on the will in harnessing society's energy. These positions were never stable or exclusive and were, as we shall see, riven by contradictions. Indeed, people on both sides of the argument frequently turned to energy's unique status as a privileged ethical standard for society—only to confront the problem that energy only had value when it was directed toward an end.

The Mechanistic Conception of Energy

During the First International Congress of Monists in Hamburg in 1911, Loeb explained "the mechanistic conception of life" to an audience of four thousand. Recounting his decade-long research into plant and animal tropism, Loeb put forward a controversial proposition: the "metaphysical term 'will'" could be replaced by the phrase "photochemical action of light."[17] As he went on to surmise, his experiments in heliotropism, chemotropism, geotropism, and galvanotropism proved that the organism could be directly controlled by modifications in the surrounding environment.[18] Loeb had come to this conclusion following research at the University at Würzburg in 1886, under the tutelage of the botanist Julius von Sachs.[19] Having begun research on psychophysics, a field fronted by the psychologist Gustav Fechner, Loeb had turned to Sachs's experiments in plant tropism to explore nondualistic accounts of the mind proposed by this emerging field.[20] Taking a question posed by Sachs—how does a seedling orient itself to the light—Loeb explored whether such basic instincts could be found in lower organisms.[21] Loeb described how lower organisms exhibited positive and negative heliotropic tendencies, like the plant. If the light that shone on a symmetrical organism was uneven, the organism readjusted its movement to correct this imbalance.[22] In some organisms this tendency was so strong that they would be compelled to follow the light, even if this would threaten their life. This phenomenon, Loeb believed, was reducible to the Bunsen-Roscoe law of reciprocity. Taking the blackening effect of light on a photographical plate as an example, the law formalized the relationship between photochemical effects, the concentration of light, and its duration. When the intensity of light doubled, the length of exposure could be halved to gain the same effect. The same law could be applied to more complex organisms, making them, in essence, "photometric machines."[23]

Inspired by Loeb's findings, the famous engineer John Hays Hammond Jr. constructed a heliotropic dog, which followed the light like Loeb's heliotropic organism. With selenium eyes responsive to light, the inclusion of a wooden nose caused light to fall on separate eyes—the imbalance forcing a process of reorientation.[24] Hammond extended this logic to produce a heliotropic torpedo, attracted to enemy searchlights, an invention that was later popularized in the high-grossing Pathé film *The New Exploits of Elaine* (1915). Hammond's heliotropic dog and torpedo simplified the relationship between light and the will to a direct mechanical response. Although this relationship was harder to trace the more complex the organism became, Loeb concluded that, due to the measurable equivalence, it was plausible to assume that "the apparent will or instinct of these animals resolves itself into a modification of the action of the muscles through the influence of light."[25] He deduced from these experiments his famous notion that the metaphysical category of the will could be reduced to the "photochemical action of light."

Loeb announced this before the German Monist League, explaining how it provided a new ethical outlook for society. Whereas previously, he described, the metaphysician determined the ethical system, now "the instincts" could be taken as "the root of our ethics."[26] "Our wishes and hopes, disappointments and sufferings," he declared, "have their source in instincts which are comparable to the light instinct of the heliotropic animals":[27] "We eat, drink, and reproduce not because mankind has reached an agreement that this is desirable, but because, machine-like, we are compelled to do so. . . . Economic, social, and political conditions or ignorance and superstition may warp and inhibit the inherited instincts and thus create a civilization with a faulty or low development of ethics."[28] Both

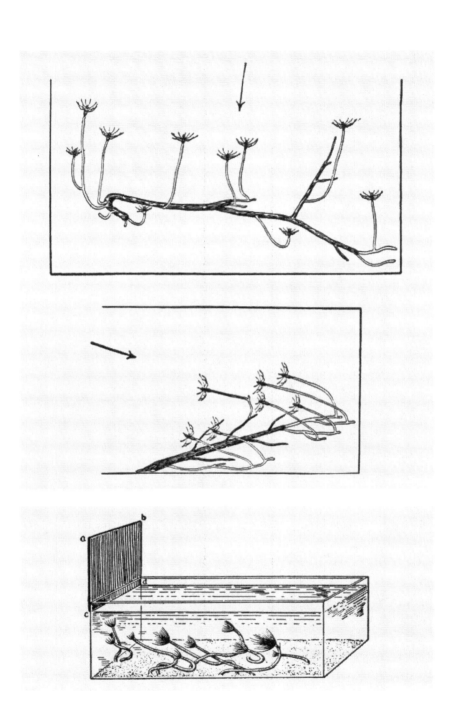

Illustration of positive heliotropism in polyps of Eudendrium and marine worms (Spirographie). The organisms bend toward the source of light, which is indicated by an arrow. J. Loeb, *The Mechanistic Conception of Life*, 28, 29.

his talk and his book *The Mechanistic Conception of Life* (1912), published a year later, caused a stir in the press.[29] One journalist in *Metropolitan Magazine* asked whether Loeb had provided the chemistry for the phrase "I am the Light of the World." With all things considered, he continued, Loeb's experiments proved that "the Will of Man is no freer than a moth."[30] Another commentator writing from Tennessee insinuated that heliotropism could explain the human attraction to taverns. If men were positively heliotropic, he postulated, surely the "dim religious light" of church would repel the subject, whereas "the gleaming plate glass, the shining brass and the glowing globes of light draw him irresistibly to the spot where, as it happens, many kinds of star-spangled drinks are served!"[31] Just as some attributed social problems, such as alcoholism, to this positive heliotropism, others fretted about the impact it would have on morality. The *Brooklyn Citizen* carried an article titled "Morality More Than an Instinct," which reported a debate on the subject held at the Brooklyn Society of Ethical Culture. How could it be, the question was raised, "that the child who steals a lump of sugar is no more guilty of sin than the bee that flies without any choice on its part towards the nectar in a flower"?[32]

As media reports considered the threat Loeb's thesis posed to society, the science journalist Carl Snyder, writing for the *New York Times*, ran with the headline "Theory of Life: Dr. Loeb's Mechanistic Conception of Energy."[33] Pointing out that there had been "no more revolutionary work" written "since the establishment of the Copernican theory," Snyder argued that "fifty years hence" this volume should "be recognized as the foundation of a new moral philosophy."[34] That this "mechanistic conception of energy" posed a threat to moral philosophy as radical as the Copernican theory illustrated the fact that energy had become as much a matter of ethical debate as one of scientific debate. This was a reference to the long-standing mechanist and vitalist controversy that centered on whether energy was purely physicochemical or consisted of some additional unaccounted-for element. During the years in which Loeb was writing, examples of this vitalist energy could be found everywhere: in Henri Bergson's élan vital, in Hans Driesch's "entelechy," in George Bernard Shaw's "Life Force," and in Jan Christiaan Smuts's "evolutionary holism."[35] Recounting the progress wrought by Loeb's discovery in the avant-garde journal the *Dial*, the biologist Raymond Pearl outlined the debate.[36] He laid out the three possible determinations and frameworks to understand this problem: either

> Life = Inorganic matter + physico-chemical energy and forces + *nothing more whatsoever*,

which is the mechanistic conception of life; or
> Life = Inorganic matter + physico-chemical energy and forces + a special *vital* element different in kind, and *in toto*, from everything not living;

which is the vitalist's rendering of the equation of life; or, finally,
> Life = Inorganic matter + physico-chemical energy and forces + *(f)(x)*, where *(x)* denotes the great *unknown* beyond the boundaries of the meagre store of knowledge of life which biological science has so far accumulated,

which is the position of the rational agnostic, who simply believes that the present state of knowledge does not warrant him in definitely asserting that *x* is what either his vitalistic or his mechanistic friends say it is.[37]

Although Pearl argued that Loeb was "the most ardent and uncompromising of living advocates" of the first point of view; he reasoned that "nobody settles the question by making up his mind."[38]

The Monist Worldview

The German Monist League, being a heterogeneous organization that incorporated a wide range of perspectives, also had to confront what sort of energy best suited the monist worldview.[39] Founded in 1906 by the German biologist Ernst Haeckel, the League became an influential force in Germany, calling for a new social, cultural, religious, and political foundation for German society, based on the fundamental unity promised by monism.[40] This included the promotion of a cooperative state that rejected liberalism and democracy in the name of an integrated society in which members saw themselves as part of a continuous whole, working toward the common good.[41] This social ideal informed the German chemist Wilhelm Ostwald's program of "energetics," and he used the German Monist League—acting as its president from 1911 to 1915—to promote a range of practical programs that applied the logic of energy to society. Developed originally as a science, Ostwald's energetics stressed that all matter (including atoms, molecules, and ions) could be subordinated to energy—matter being a fiction that helped the rational subject comprehend the perpetual transformation of energy. Ostwald, however, moved away from this reductionist view of energy as an essential reality, to posit the law of "efficiency" as the fundamental principle of monism.[42] Ostwald's monism rested on the premise that all the actions of humankind—from technology, to everyday habits, to the "highest sociological and ethical problems"—are constructed under the impulse to reduce the amount of energy spent.[43] In this way, he transformed energy into a moral law, "the en-

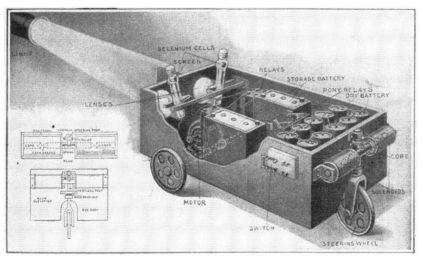

Top, The electric dog invented by John Hays Hammond Jr., being controlled by a flashlight. *Bottom*, The internal mechanism of the electric dog, accompanied by a diagram. B. F. Meissner, "The Electric Dog and How He Obeys His Flashlamp Master," *Popular Science* 88, no. 3 (March 1916): 426, 429.

ergetic imperative," based on the maxim *"Waste no energy; turn it all to account."*[44]

Ostwald suggested a range of practical applications for the "energetic imperative," the most extensive being "Die Brücke" (the Bridge, International Institute for the Organisation of Intellectual Work). The Bridge sought to establish a

"World Brain" by standardizing models of communication, creating a universal language (based on Esperanto), a world encyclopedia, and a "World Format" for publication standards.[45] Recognizing the current drive toward collaboration and standardization in international life—in areas such as the global economy—Ostwald hoped that his "Brain Trust" would do the same for intellectual life. It would enable "the total work [to] be so organized and brought into harmonious union that no energy is wasted."[46]

In fact, during the 1911 conference at which Loeb gave his address, Ostwald had drawn on the principle of energy to call for a "world organization" to apply the methods of "technocratic management" to the scattered efforts of scientists.[47] Here the conservation of energy operated as a model to standardize the world's information, to regularize it into a common standard that could be translated between national borders, as effortlessly as when energy was translated into different forms. Not only, however, was his energetic imperative an instrumental principle, but he also went on to explain how, as a fundamental monism, energy drove the foundation of the highest social forms. This included "the constitutional state" (which reduced conflict among individuals), pacifism (which stopped energy from being wasted in warfare), and the collectivist state (the "perception that the individual is more and more a mere cell in the collective organism of humanity").[48] This monism, Ostwald believed, could even surpass the highest ideals of Christianity. Once the energetic imperative was taken to heart, "the evolution of kindness and love, the evolution of the spirit of self-sacrifice and devotion to the great whole of humanity" as the basis of Christian ethics, could become "more and more a demand of the energetic imperative, therefore an immanent demand of our whole Monistically ordered life."[49]

Despite his leading role in the organization, Ostwald's energetics was not the dominant viewpoint. In fact, it had been subject to critique, ever since it was attacked by Ludwig Boltzmann in 1895 at the Lübeck *Naturforscherversammlung*.[50] Instead, although Ostwald recognized that energy was essential to a monist worldview, how energy supported this paradigm remained up for debate. One of the central controversies rested on energy's relationship to the will. In 1908 the *Literary Digest* dedicated an article to this ongoing dispute, recounting a recent controversy over the subject that had erupted in the American philosophical periodical the *Monist*.[51] Edited by Paul Carus, the *Monist* was central to promoting a monist worldview in the United States.[52] It had carried Ernst Mach's and Haeckel's writing, as well as providing information on Ostwald's energetics and Loeb's mechanistic science. The debate centered on the independence of the will and rehearsed long-held positions as to whether the will was reducible to energy laws or autonomous from them. The first article, written by A. E. A. Wilkinson, put forward a case for free will, arguing that consciousness does, "ever and again, originate energy."[53] This followed an ancillary logic, whereby spiritual energy (emanating endogenously from the subject) increased in the universe, rather than diminishing according to the laws of entropy. "Will-force," he argued, upset the trend toward equilibrium, as "the total supply of spiritual force in the universe must be infinite; otherwise it would be liable to exhaustion."[54] The second article, by William Pepperrell Montague, took a determinist perspective, arguing that since mental processes occur in space, material energy was transferable into mental energy, upsetting the mind-body dualism.[55] The third article, by Carus, disputed both these views. Since the monist perspective required continuity between all fields, he argued that the will does not "create energy" but redirects it within the cosmic order. The will imparts direction to the energy already present, but "the position of the ship's rudder is a state or condition," and "to adjust the rudder according to requirements takes a certain, albeit a comparatively small, amount of energy."[56]

Each of these positions built on nineteenth-century debates as to whether energy strengthened or undermined free will. The belief that the will was a form of energy had a long genealogy within Western thought. The German philosopher Arthur Schopenhauer had notably argued in his *The World as Will and Representation* (1819) that all activity and energy in nature were reducible to the will. This supposition had been strengthened by the discovery of thermodynamics, which bound the will into the strict determinism of the energy laws. By applying the conservation principle to biological phenomena, Hermann von Helmholtz undermined any autonomous vital principle operating outside of the energy laws.[57] This rigid determinism was highlighted by John Tyndall during his 1874 "Belfast Address," given at the British Association for the Advancement of Science, where he declared the entire domain of human knowledge "fast in fate" to the conservation of energy.[58] While the conservation laws presupposed the eradication of free will from the cosmic scheme, those unwilling to give up on volition reinscribed it, attributing to it an extraphysical agency, or relaxing the determinism of the energy laws altogether.[59] In fact, the historical roots of Maxwell's demon, as Crosbie Smith and M. Norton Wise have demonstrated, were embedded in a defense of free will against the rigid determinism posed by Tyndall and Helmholtz.[60] Able to reverse the process of entropy by sorting atoms, Maxwell's demon introduced statistical indeterminacy into the second law and thereby allowed room for volition. By conjuring up a demon that resembled "a doorkeeper very intelligent and exceedingly quick," Maxwell reinscribed the faculty of the will as it redirected energy within the universe.[61] As this controversy rolled on, the following questions remained: Was the will a sorting demon redirecting energy through the aid of instruments, "cricket bats, valves, arms, trap doors, or switches"?[62] Did the will presuppose countless sources of individualized consciousness self-generating energy? Or was the will bound by the determinism of the energy laws? The challenge went both ways: just as the possibility of free will threatened the principle of conservation, the laws of thermodynamics undermined any scope for free will.[63]

Weighing in on this debate was the Viennese positivist Ernst Mach.[64] In two studies, *Principles of the Theory of Heat* (1886) and *History and Root of the Principle of the Conservation of Energy* (1911), Mach had argued that energy not only was a physical constitution of matter but also existed as a means of description.[65] Providing a history of scientific explanations of energy and tracing how they moved toward greater precision, Mach concluded that the laws of thermodynamics were the most efficient "economy of thought," an articulate scheme for mapping relations.[66] Just as labor relations were described through economic ordering, intellectual activities, Mach maintained, demonstrated a similar economy, making the laws of thermodynamics the most "economic" paradigm for ordering phenomena within a particular economy. Within this "economy of thought," an isolated entity, such as the will, outside causation was unfeasible in the same way as perpetual motion remained an impossibility.

Ostwald's energetics developed out of Mach's positivism.[67] Just as Mach posited the laws of thermodynamics as an "economy of thought," an ordering system within the cosmos, Ostwald took energy to be the optimal way to understand all forms of experience in a uniform manner.[68] Within this Machian "economy of thought," boundaries between the mind and the body were dissolved once energy became the unifying logic. In 1906 Ostwald had expanded this thesis in a talk given at the American Psychological Association on the subject of "Psychical Energy," where he argued that all "psychic facts," like physical ones, could be measured as a quantity of energy, making the study of psychology reducible to energetics.[69] Although rebuttals came from figures like the philosopher Josiah Royce, who raised doubts about whether mental phenomena could be measured as a physical energy, Ostwald was adamant that once it was evident how much energy there was in the body at any one moment (by calculating how much energy did not appear in another form of work) one would be able to quantify the amount of psychical energy utilized in a mental act.[70]

To Ostwald, however, the ability to measure all phenomena—including thought—as a magnitude of energy was as much a social promise as a scientific one.[71] He outlined the ethical implications of this viewpoint in a talk titled "Individuality and Immortality," given at Harvard University the same year.[72] In the talk, Ostwald explained how energy laid the foundation of a new ethics, based on the principle that the logic of "diffusion, or a homogenous distribution, of energy is the general aim of all happenings."[73] He described how "no change whatever seems to have occurred, and probably none ever will occur, resulting in a concentration greater than the corresponding dissipation of energy."[74] Because of the natural tendency toward diffusion rather than distinctiveness, Ostwald posited the highest forms of culture to be those that moved in this direction—forms that "diminish the differences between men. It equalizes not only the general standards of living, but attenuates also even the natural differences of sex and age."[75] In opposition, moments of higher concentration and "individuality," found, for instance, "upon the accumulation of enormous wealth in the hands of a single man," stood as "an imperfect state of culture."[76] Mirroring this correlation, moments when the "will" was lost—in love, or in the appreciation of art or music, "the happiest moments of our life"—occur when energy follows its natural course and one forgets oneself "as a drop is carried by a wave."[77] These were the moments when the boundary between inside and outside was eradicated and the homogeneous distribution of energy upset social atomization. In Ostwald's view, the state of "individuality" was equivalent to "limitations and unhappiness," whereas the natural tendency of energy toward diffusion and homogeneous distribution prescribed an ethics based on collective organization.[78] Once the individual was taken to be part of an integrated whole, part of a single organism or energy system, the less, Ostwald believed, would he be able to separate his own personal interests from the common good.[79]

Jacques Loeb's Energy Ethics

Loeb paid homage to both Ostwald's and Mach's epistemological models in an article published in the *Yale Review* in 1915 titled "Mechanistic Science and Metaphysical Romance." Along with Ostwald, Loeb counted Mach his "intellectual mentor."[80] Following his address at the Monist League, Loeb had actively worked for the cause of monism in the United States, even trying to find some reconciliation between the philosophy of monism and the political aspects of social democracy.[81] However, he rejected the more experimental methods proposed by Ostwald, such as founding a monist village in the United States, and as the war broke out Loeb gradually distanced himself from Ostwald and the Monist League (which collapsed in 1915).[82] Nonetheless, despite this break, Loeb paid tribute to Ostwald and Mach in the article, pointing out that, despite the fact that he rejected their scientific positions, these thinkers had sought a complete description of nature and thereby had developed an important social and ethical vision.

In the article, he described metaphysical thinking as a form of dangerous Romantic speculation, making it an "unbalanced system."[83] "A good deal of this 'philosophy,'" Loeb even hazarded in a draft, seemed to be "tainted by mental derangement."[84] Set outside a system of quantitative relations, it could be found in the language of the "German nationalist" aroused by "Nietzsche," the "English nationalists" incited by "Kipling," and the reactionary element among Bergson's "French patriots."[85] By following an unbalanced energetics, these forms caused disjunction in the system. Through unhealthy acts of concentration, or, as Ostwald had argued, the "accumulation of enormous wealth in the hands of a single man," a continuous energy economy would be individualized into heterogeneous, individualized, unbalanced centers. In this respect, whether "humanity" was guided by "mechanistic science or metaphysical romance" was not just an "academic" problem. It informed competing ethical systems.

To counter metaphysical abstractions like the will, Loeb's mechanistic conception of energy provided a model for social reformers, intellectuals, and writers of a fairer society grounded in energy laws, in a homogeneous energy economy rather than one that depended on individualized heterogeneous centers. His biological program also came to resemble the social critique put forward by the economist Thorstein Veblen.[86] This similarity was hardly surprising, as Veblen and Loeb developed their thought in tandem. They had been colleagues at the University of Chicago in the 1890s, and once Loeb left Chicago in 1902 for the University of California, Berkeley, they remained in contact, corresponding throughout their careers.[87] Veblen had gone farthest in applying the scientific method to the technical infrastructure of the state and its political economy. In his canonical work *The Theory of the Leisure Class* (1899), Veblen had described how pecuniary culture siphoned energy away from peaceful society toward the private economies of the individual and the corporation. Veblen argued that the dictates of "conspicuous consumption" required surplus energy and waste, whereby the leisure class withdraws the necessity of daily subsistence from the bottom: "reducing their consumption, and consequently their available energy."[88] In order to do this, predatory culture, Veblen recognized, had constructed all sorts of social rewards that justified "all manifestations of force in terms of personality or 'will power.'"[89] For example, he explained how the model of "barbarian prowess" at the heart of predatory culture depended on two main directives: "force and fraud."[90] It relied on an animalistic mindset and, above all, the dictates of luck. The provided epistemological framework gave the illusion of operating outside biological and economic limits of causality. It drew on an extraphysical, unbalanced, vitalist force.

In his later book *The Instinct of Workmanship, and the State of the Industrial Arts* (1914), Veblen described Henri Bergson's élan vital as a relic of this primitive mindset, its "creative bent" indistinguishable from "the *mana* of the Melanesians ... the *wakonda* of the Sioux ... or even the *hamingia* of Scandinavian paganism."[91] Along with "magic, occult science, telepathy, spiritualism, vitalism, pragmatism," its epistemological structure did not match the sensibility of the machine age—it was out of date.[92] In contrast, the conservation of energy—even though a "metaphysical postulate" itself—was an epistemological tool that had found a home in the age of the technological machine.[93] Calculable, quantifiable, and able to be balanced, mechanistic energy, Veblen believed, had to be extended beyond the sciences, to supplant the animalistic, barbarian mindset that still dominated the nation's social institutions, customs, and laws.

Alongside Veblen, intellectuals, writers, and political activists would be drawn to Loeb's mechanistic conception of energy as a way of eliminating the vitalist elements that drove pecuniary culture. Loeb's mechanistic science would inspire not only writers such as Theodore Dreiser, Upton Sinclair, and Sinclair Lewis but also religious leaders, who identified it as a blueprint for a better society.[94] For instance, George Willis Cooke, a Unitarian minister, campaigned for Loeb's energy conception to become the basis of a new socialism.[95] The new mysticism present in ontologies like George Bernard Shaw's "Life Force," Cooke believed, covered up "old prejudices, credulities, sectarianisms, and superstitions."[96] Despite Shaw's "Life Force" being associated with a socialist agenda, Cooke pointed out, it was merely "old priest writ large." Shaw's "Life Force" collapsed a Nietzschean ontology with a powerful energy drive to center human volition, physical strength, and intellectual strength at the heart of a socialist Fabianism.[97] Following the neo-Lamarckian thesis of recapitulation, this "Life Force" found its perfect expression in an ideal "Superman," as energy emerged from the "will of the subject," stored in memory through countless generations. While Shaw's "Life Force" relied on an endogenous energy within the subject, in Loeb's mechanistic science energy would be balanced within a wider economy. Rather than the subject surpassing his energy revenue through a predatory force, mechanistic energy spread responsibility through the system.

Energy and Social Evolution (or Devolution)

Energy, as it eliminated the will, provided a powerful ethical model for society grounded in mechanistic science to counter the militarism and capitalist exploitation endemic to the industrial age. In the hands of other prominent intellectuals, however, subordinating the will to energy could also support the very ideologies Loeb's mechanistic science sought to counter. Social Darwinism, for example, which celebrated the capacity of the strong to triumph over the weak, also found a powerful defense in energy and its ability to eliminate the will from the social order.

One particularly illuminating example of how energy was used to promote the ideology of social Darwinism was through the work of Henry Fairfield Osborn, president of the American Museum of Natural History, who published *The Origin and Evolution of Life: On the Theory of Action, Reaction and Interaction of Energy* in 1917. Born into a wealthy railway family and a powerful member of the New York elite, Osborn drew on a descriptive science riddled with the rhetoric of race hatred and social Darwinism that Loeb detested. In *The Origin and Evolution of Life*, Osborn combined his own religious faith as a devout Christian with Loeb's physicochemical outlook to posit an energy that flowed toward a predestined end. In the book, Osborn described how, through the dynamics of "action, reaction and interaction," evolution could be explained as the product of constantly developing energy assemblages.[98] This energy conception of life relied on the interaction of four energy complexes: the inorganic environment, the developing organism, the germ or hereditary chromatin, and the life environment. This model, which he termed the "energy concept of life," replaced the privilege of "form" in Darwin's evolution with Loeb's concept of physicochemical energy. He wrote,

> The evolution of life may be rewritten in terms of invisible energy, as it has long since been written in terms of visible form. All visible tissues, organs, and structures are seen to be the more or less simple or elaborate agents of the different modes of energy. One after another special groups of tissues and organs are created and coordinated—organs for the *capture* of energy from the inorganic environment and from the life environment, organs for the *storage* of energy, organs for the *transformation* of energy from the potential state into the states of motion and heat. Other agents of control are evolved to bring about a harmonious *balance* between the various organs and tissues in which energy is *released*, hastened or *accelerated*, slowed down or *retarded*, or actually arrested or *inhibited*.[99]

In Osborn's energy conception, selection was not an energy complex but an arbiter between different complexes and forms of energy: "it antedates the origin of life."[100] Osborn's "energy conception" thus rooted a creative force at the heart of the cosmic process, while leaving room for the will as part of the evolutionary struggle. Osborn's former mentor, the American paleontologist Edward Drinker Cope, had introduced a similar generative energy to describe growth processes, which he termed "bathmism," or "bathmic energy."[101] This force was a form of "intelligent direction"—a supraconsciousness—guiding the evolutionary process to perfect form.[102] Replacing Cope's "bathmic energy" with Loeb's physicochemical laws, Osborn maintained that his model denied any "supernatural or teleological imposition through an externally creative power."[103] For Osborn it was possible to rule out any unscientific theory of "an *entelechy* or any other form of internal perfecting agency distinct from known or unknown physiochemical energies."[104] Still, he reinstated Cope's "bathmic energy," claiming that the law of progress was immanent to the "energy conception" driving the evolutionary process forward to a predetermined end. One reviewer noted in *Current Opinion* that "energy appears to be the chief end of life."[105] In comparison, "form is relatively simple . . . the lowest organisms seem to be almost formless, while energy is everything."[106]

Loeb remained unconvinced by Osborn's unscientific application of the energy laws, as did the geneticist Thomas Hunt Morgan, who worked with Loeb to promote positivist science in the United States.[107] Sending Loeb a copy of a letter forwarded to Osborn, Morgan pointed to the many errors present in Osborn's energy conception.[108] The letter praised Osborn's ambition in explaining evolution through the physical laws of energy.[109] And yet, Morgan criticized Osborn for inserting a "mystification" at the heart of the matter and for stumbling into superstitious fallacies while clinging to the energy laws.[110] Prime among these was the confusion between "chance" and "law" running through Osborn's text. If Osborn wanted to adhere to the laws of thermodynamics, this dichotomy, Morgan asserted, was "nonsense."[111] To make this misjudgment work, one had to "throw over the energy conception and substitute a mysterious and beneficent being that directs all things that are lawful and introduce a devil-of-a-fellow to mess things up generally and take a chance of coming out all right at the end."[112] Taking all this into consideration, Morgan blamed Osborn for elevating certain forms of life above the rest of living matter: "even by intimating the possibility of such a conclusion you leave the safe fields of the energy conception and roam abroad in the Elysian fields of mystery."[113]

Morgan's critique of Osborn illuminates what was at stake in the application of energy laws. Not only did Osborn's manipulation of the energy laws violate the empirical science Loeb and Morgan were advocating, but it pointed toward an entirely antithetical social order.[114] The implications of this were evident in the high praise bestowed on the book by Theodore Roosevelt, who compared it to Huxley's and Darwin's work in importance. He complemented Osborn for "clearly grasping the fact that *energy*, and not *form*, lies at the beginning of the evolution of life."[115] Thanks to Osborn, he explained, "it seems to be clearly proven that life evolves in an orderly way; and this is one reason for believing that the energy which keeps the universe in order is, in some way which we do not comprehend, also responsible for the orderly procedure of life."[116] It is hardly surprising that Roosevelt seized on Osborn's progressive energetics. Not only had Osborn placed energy at the heart of evolution, but he also reinscribed a predetermined order into Loeb's chance-based energy, which would have pleased a fanatical advocate of progress like Roosevelt. It did not take too wild a leap of the imagination to read Osborn's energy conception as a grand allegory of the American nation, with those most able to store, capture, and accelerate energy leading the nation forward toward its manifest destiny. Even though this was not explicit in the text, this was how Osborn's readers responded to the book, praising him for unraveling a political, social, industrial, religious, and aesthetic law for the nation, grounded in energy.[117] On reading the book, Roosevelt may have even flattered himself that his friend (and fellow member of the Boone and Crockett Club of Manhattan) was thinking of him, the Rough Rider, when he described Darwinian struggle through the energy laws.[118] Roosevelt, after all, had exhibited a prodigal energy-wielding capacity during many a high-profile stunt: trekking through deadly blizzards, boxing prime fighters, and hunting wild game in Africa, to name the more apocryphal stories. Vanity aside, what was of most significance was that, in placing the law of "action, reaction and interaction" at the heart of his energy conception, Osborn had provided Roosevelt with a model of energy that affirmed, rather than undermined, the active role of the will in the cosmic process.

This would have additional importance for Roosevelt. As part of a fin de siècle generation obsessed with the repercussions of the second law, he had spent the best part of two decades (as both president and national celebrity) promoting the will as a mechanism to overcome the gradual siphoning off of energy. By the end of the nineteenth century, conflicts between the devolutionists, whipped up by Max Nordau's book *Degeneration* (1892), and social Darwinists (led by Herbert Spencer) had called into question how far the energy laws could be extended and whether the second law undermined evolutionary progress, or vice versa.[119] This fear had been encapsulated by the popular medical malady known as neurasthenia. Neurasthenia had been first described by the physician George Miller Beard in the 1860s and was later outlined in his studies *A Practical Treatise on Nervous Exhaustion, (Neurasthenia); Its Symptoms, Nature, Sequences, Treatment* (1880) and *American Nervousness: Its Causes and Consequences* (1881).[120] The disease referred to a weakness of will—a "nervous bankruptcy"—brought about by the draining of "nerve force," and it came with a range of symptoms, including sleeplessness, anxiety, body pains, and headaches.[121] The disease would be credited to the overstimulation inherent to modern life, a result of the acceleration brought about by the steam engine, electricity, new modes of communication, and the modern press. Combine this with the *"tedium vitae,"* the rapid bureaucratization of all spheres of life and the emergence of white-collar jobs, and you had an ailment characterized by inertia, as people found themselves unable to act vigorously in the face of industrial society. By the turn of the century, neurasthenia had become an upper-class obsession, charged with having made an entire generation "as defunct as the dodo . . . just jellyfish, and flabby all through."[122]

More importantly, neurasthenia served a symbolic function, mirroring a cultural fixation with the repercussions of the second law for the nation.[123] Here the subject's energy economy, imagined by Beard as a "battery" or a bank "account," was equated to the "social organism," as both were subject to entropy.[124] This neurosis saturated the cultural imaginary. It was implicit, for example, within Frederick Jackson Turner's frontier thesis, which drew attention to the physical limitations imposed on the nation's natural resources. At the heart of the cultural fascination with the second law, however, was the problem of the will as people confronted the dilemma: was the will—as the logic of neurasthenia suggested—subject to the second law, or could the will, as many therapeutic remedies promised, thwart its logic? Could the will, in essence, become a center of energy?[125] Henry Adams and Brooks Adams, brothers and heirs of the Adams political dynasty, would confront this problem in different ways, as they came to apply the energy laws to the study of history.

Brooks Adams, the younger brother by ten years, was the first to address this subject in *The Law of Civilization and Decay* (1895). In this book he provided a scientific "Law" for society as it oscillated between "barbarism and civilization," or, in his own words, as human societies move from "physical dispersion to one of concentration."[126] For Brooks Adams, the progress of civilization depended on the instincts through which energy was channeled. This had led to three different levels of society: the imaginative, the martial, and the economic.[127] These stages reflected changes in the social structure, progressing from scattered primitive societies, where energy was guided through "Fear," toward imaginative pursuits, to centralized states dominated by "Greed," in which energy was dissipated "in war and trade."[128] As scattered communities became increasingly concentrated, energy no longer expended itself in daily struggle but was stored as wealth.[129] In the final stage of centralization, arising within an economic society, physical energy was transformed into money, leading to its control by an autocratic minority and the "waste of energetic material" rather than its redistribution through society.[130] At this point, society reached a pressure point leading to either stasis or a disintegration that led back to barbarism.

It is easy to guess what stage in the process Brooks Adams posited contemporary America to be in. Furious over the 1893 economic panic, which had almost led the Adams family to bankruptcy, Brooks Adams (who, like his brother, idolized the virtues of old-school Republicanism) interpreted the sleazy commercialism of contemporary society as a Jewish conspiracy, with the banks in control of the world's energy. All around him, Brooks Adams found evidence that fin de siècle society was on the brink of collapse—with "the architecture, the sculpture, and the coinage" at the close of the nineteenth century resembling the "Rome of Caracalla" rather than the "Athens of Pericles."[131]

Brooks Adams's "Law" was cyclical: energy would dissipate, but it would be recouped as society fluctuated between concentration and dissipation. Writing over ten years later, however, Henry Adams took a more pessimistic view, turning to the second law to project a linear model that embraced the entire span of history as a direct descent toward a terminus point. Adams had come to the second law after applying the rule of force to history in his two earlier studies, *Mont-Saint-Michel and Chartres* (1904) and *The Education of Henry Adams* (1907). However, whereas the virgin in these earlier studies came to stand for the importance of will in imparting direction to energy, in his later study (and most literal exploration of this subject) *A Letter to American Teachers of History* (1910) Adams turned to the work of Loeb and Ostwald to reduce the will to energy.[132] He described how Loeb's experiments in animal tropism had presented a challenge to the entire philosophical order by reducing the will to electricity and chemical affinity. And yet, Adams pushed this one step further than Loeb, reducing the will to the second law.[133] As such, Adams turned evolution on its head. Noting how historians had taken reason and the intellect as the highest achievement of man, the "highest energy of nature," Adams maintained that the will (as revealed in the tropic response) was the highest form of "vital Energy." Consciousness, thought, and reason, on the other hand, were the product of "lower energies," a sign of the gradual dissipation of the will. As such, the highest energy, Adams believed, was found rather in the "Scarab or the Scorpion, where it is unconscious, than in Monkey or Man, where it is conscious."[134] Because of this reversal, Adams had to conclude that "if all the other sciences affirm that not Thought but Instinct is the potential of Vital Energy ... nothing remains for the historian to describe or develop except the history of a more or less mechanical dissolution."[135]

The Adams brothers thus applied the energy laws to different effect.[136] Where Henry Adams saw history as a linear process of mechanical dissolution, Brooks Adams remained optimistic—by intervening in its process, the "Law" could be slowed down. Brooks Adams was further encouraged by Theodore Roosevelt, who, as he wrote in a review published in the *Forum*, felt that his friend had ignored many of the current opportunities to recoup energy in contemporary society.[137] Brooks Adams later became a confidant to and strategist for an inner circle of Washington grandees, having the ear of Roosevelt, along with other powerful notables like Cushman Davis, chairman of the Foreign Relations Committee, and Henry Cabot Lodge (who made sure that every Supreme Court justice received a copy of the "Law"). In an influential position, Brooks Adams advocated many solutions to counteract the inevitability of the "Law," placing the onus on a strong will to slow its advance.[138] This led Brooks Adams down a paradoxical line of thought in which he supported the very processes he had fought against: increased centralization, control, and expansion. His program to fight the "Law," therefore, mimicked its internal dynamics as he proposed a range of measures to thwart its progress: the United States had to expand its administrative mechanism to control the surplus energy wasted through unregulated markets; it would become an imperial force maintaining dominion over central trade routes, in order to gain access to key reserves of potential energy; and finally, it had to reinvigorate the "Fear" instinct so that "martial man" could supplant economic man, whose "Greed" instinct led to the dissipation of energy.[139]

Over the next two decades, Brooks Adams would not waver over the validity of the "Law." Instead, he came to echo progressive rhetoric that called for technocratic government, greater efficiency, and the minimization of drift.[140] Moreover, during this period he transformed the "Law" into a grand political synthesis that promoted the state as an ideal vessel, a comprehensive will, to control the chaotic energies generated within a capitalist democracy. Democratic society required a strong mechanism to control the self-interested competitive urges driven by the market and by the "Greed" instinct of man; it needed "a centre of energy whose will must dominate."[141] The genius of George Washington was emblematic of this strong will; his power lay in the control he exerted over the great mass of energy within the North American continent. In Brooks Adams's view, Washington recognized that governments worked as a "collective brain" whose chief operation was to extend a powerful control over the nation's energy.[142] Since George Washington's death, this overriding will had rotted away, leading by 1828 to a level of "degradation" that matched the fall in "intelligence and intellectual energy" of the political state under Andrew Jackson.[143] According to the "Law," the

"Greed" instinct that predominated under Andrew Jackson revealed the fault lines running through modern democracy. It was, in Brooks Adams's words, "an infinite mass of conflicting minds and of conflicting interests which, by the persistent action of such a solvent as the modern or competitive industrial system, becomes resolved into what is, in substance, a vapor, which loses in collective intellectual energy in proportion to the perfection of its expansion."[144]

By 1915, debates over the relationship between energy and the will had coalesced into relatively distinct positions, each with their own social and political implications. For those who promoted a monist worldview, such as Loeb, energy promised a continuous economy in which metaphysical categories, such as the will, and unhealthy states of "individualization" would be eliminated. In the hands of social Darwinists like Osborn, energy was inscribed with a supraconscious will, which led evolution to a predetermined end. Within Brooks Adams's political theory, the will remained central to social progress, providing support for greater centralization of the state. For Henry Adams, who applied the second law of thermodynamics to the will, contemporary society was reaching a terminus as the energy of the will dissipated to give way to thought and consciousness.

William James's Dynamogenic Will Power

One final position promoted the will as a great liberator of energy. It was the Harvard psychologist and public philosopher William James who did the most to cement this point of view. James had a personal stake in the relationship between energy and the will, having spent periods of his adult life bedridden with crippling neurasthenia. His pragmatic philosophy, in this respect, could be seen as a response to the neurasthenic torpor of his generation and his own failure of nerve.[145] Throughout his philosophical program, James placed the individual will at the center of the energy system.

James was quick to critique those who adopted energy as an ethical system on its own. For example, on reading Henry Adams's *Letter*, he struggled to accept his old friend and ex-colleague's determinist adherence to the second law. In a playful letter to Adams, written in the last weeks of his life, James pointed to an obvious flaw in his "tragic subject": Adams had confused "the *amount* of cosmic energy it costs to buy a certain distribution of fact."[146] As such, although he did not deny that the second law provided a "terminus" for history, he had to remind Adams that certain elements positioned on the same *"energy-level"* were not uniform— but had a built-in hierarchy.[147] He explained: "Physically a dinosaur's brain may show as much intensity of energy-exchange as a man's, but it can do infinitely fewer things, because as a force of detent it can only unlock the dinosaur's muscles, while the man's brain, by unlocking far feebler muscles, indirectly can by their means issue proclamations, write books, describe Chartres Cathedral, etc., and guide the energies of the shrinking sun into channels which never would have been entered otherwise—in short, make history."[148] James thus took aim at Adams's emphasis on "absolute physical units" rather than considering what energy was actually used for.[149] For James, it was a fallacy to take energy as an absolute magnitude: "quantity and distribution" were not the same, even though, he stressed, it was common to assume that they were.[150]

This critique was not only reserved for Adams's clumsy application of the second law; James also identified a similar error in Herbert Spencer's model of the "persistence of force."[151] James criticized Spencer's concept of "force" for being "one vast vagueness" and for connecting postulates such as "mental force" and "social force" to energy budgets (a "quantity absorbed from nature").[152] This, he believed, was founded on a range of misconceptions. Prime among these was the assumed correlation between the "nature of *releases*" and a "quantity antecedently absorbed from physical nature."[153] "A discovery, a book, a new idea, or a national insult" could become one of the "greatest social forces," and yet, as James pointed out, this had little relation to a quantity of physical energy: "the greatest of 'forces' of this kind need embody no more 'physical force' than the smallest."[154] Even "Mr. Spencer himself," James clarified, neatly demonstrated the paradox inherent in this view: he "is a great social force; but he ate no more than an average man, and his body, if cremated, would disengage no more energy."[155] James was thus quick to critique the assumption, taken up by Spencer and Henry Adams, that energy was a direct correlate of social value. Instead, energy could not be reduced to a physical property, as its value depended on its direction as conditioned through the will. He explained as much to Adams, reminding him that "as the great irrigation-reservoir empties itself" the whole question is not the amount of energy left but "*which* rills to guide it into.... Just so of human institutions—their value has in strict theory nothing whatever to do with their energy-budget—being wholly a question of the *form* the energy flows through."[156]

In opposition to rigid determinists like Loeb, who reduced the will to energy, James's philosophy of affirmation foregrounded the will as a harnesser of energy. Consciousness, for James, was teleological in that it held an object of thought within the flux of ideas, sensations, memories, and bodily awareness according to the needs of the subject. The ability to select between simultaneous ideas—for instance, to stay in bed on a freezing morning, or to get up—required a choice, a degree of effort to hold one thought before the mind, which led to action.[157] The weakness of will that occurred in the state of abulia correlated to a lack of attention, an inability to extract from the "stream of consciousness" a stable object. In *The Will to Believe* (1896) and *The Varieties of Religious Experience* (1902) James extended this discussion, cementing the paradigm of the heroic individual able to act on the world and bring order into existence. In *Varieties*, he took the example of the saint, who, as a figure of intense religious devotion, possessed an extreme sensibility that forced him to respond to the world with explosive power. The act of conversion that the saint dramatized involved reorganizing centers of energy, as attention moved from one focus to another. For James, everybody had a *"habitual centre of his personal energy,"* and as people's attentions constantly shifted, so too did these centers: things that are "hot and vital to us to-day are cold tomorrow."[158] The act of conversion involved a shift in the center of energy so that new ideas, habits, and values were brought into the foreground of attention, into the hot centers of "dynamic energy."[159] When this process occurred, James described how "the new centre of personal energy has been subconsciously incubated so long as to be just ready to open into flower, 'hands off' is the only word for *us*, it must burst forth unaided!"[160]

The creation of new ideals, goals, values, and centers of attention, according to James, was a prerequisite for generating higher levels of energy. The projection of new possibilities enabled the heroic individual to exceed physical limitations and conjure up ever-greater levels of energy. In his presidential address to the American Philosophical Association in 1906, titled "The Energies of Men," James elucidated how this had obvious benefits for both the individual and the national economy, laying down a set of guidelines as to how one could go about harnessing this "second wind." Each individual had a habitual level of energy restricted by "fatigue." However, underneath this "imperfect vitality" there appeared "a level of new energy."[161] In fact, James suggested, "there may be layer after layer of this experience"; a "third and fourth 'wind' may supervene."[162] This energy, James explained, "offers itself as the notion of a quantity, but its ebbs and floods produce extraordinary qualitative results."[163] As quantitative forms gave way to qualitative values, the physical energy of the body had no correlate to the energy generated by the "dynamogenic" will.[164] Through activating the mind, the economic limitations of the body could be challenged: "it throws into gear energies of imagination, of will, and of mental influence over physiological processes, that usually lie dormant, and that can only be thrown into gear at all in chosen subjects."[165] "*Excitements, ideas, and effort,*" James described, were "what carry us over the dam."[166] There were, of course, limits to energy—"the trees don't grow into the sky"—but states of being dominated by love, warfare, and Yoga, along with the invigorating effect of abstract ideas and political, scientific, philosophical, and religious conversions, act dynamogenically to release their "stores of vital energy into gear."[167] Phrases such as "having a good vital tone, a high tide of spirits, an elastic temper," James claimed, implied a certain elasticity to the "mental" and "moral" work of "energy": it allowed a degree of flexibility excluded from the mechanical understanding of energy provided by physics.[168]

Nobody did more to promote James's model of the heroic saint than his former student Theodore Roosevelt.[169] At the heart of Roosevelt's model of the "strenuous life" was the heroic individual, with their capacity to harness energy. Foregrounding the will as the constructive element, Roosevelt encouraged the subject to exceed an energy budget, to outreach bodily limits. This, as many critics have recognized, was a self-conscious act, as Roosevelt connected his political agenda, both imperial and economic, to a paradigm of will power and rugged masculinity.[170] Roosevelt's bombastic rhetoric, of which he was a living example, thus worked to transform energy into an ethical system in which magnitude—the quantity of energy harnessed—was transferred into a correlate of social value.[171] For example, Roosevelt tirelessly repeated the notion that "in the abounding energy and intensity of existence in our mighty democratic Republic there is small space indeed for the idler, for the luxury-loving man who prizes ease more than hard, triumph-crowned effort."[172] Roosevelt's many pronouncements did not just work rhetorically. As Sarah Watts has demonstrated, they exerted a strong emotional hold on an uncertain public longing for a new standard of national self-fashioning, found in the archetypes of masculinity, martial strength, and individualism promoted by his philosophy of force.[173]

Through bold rhetoric, high-profile stunts, and bravado, Roosevelt elevated energy to an ethical principle that depended on magnitude, on exceeding the physical limitations imposed by the body. From this perspective, Ostwald's edict against "individuality" was counterpoised with a model of energy that supported the boundaries of the self, as the individual became an isolated center whose energy was governed by a dynamogenic will. This, of course, was subject to the same conceptual fallacy that led Ostwald and Henry Adams to bestow ethical advantage on energy as a social principle. Energy—when tied to the symbolism of "toil and strife"—was given its own moral priority as it was converted into an ethical system based on magnitude. And yet, rather than providing an ethics that opposed "individualism," when put to work within the rhetoric of the "strenuous life," this reinterpretation of the relationship between energy and the will supported the opposite values, reinforcing the qualities of aggressive militarism, imperialism, and capitalism.

James had long recognized this shortfall in Roosevelt's jingoistic espousal of the strenuous life and subjected it to the same criticism that he later launched against Adams and Spencer.[174] This only grew more apparent once James, a member of the American Anti-Imperialist League and a pacifist, watched Roosevelt use this paradigm of the heroic life to justify America's policy of imperial expansion. When Roosevelt introduced his famous espousal of a "life of toil and effort, of labor and strife" during his 1899 address, given in April 10 at the Hamilton Club in Chicago, he was directly addressing foreign policy, calling for the United States to build up its army to assert greater imperial power in Cuba, Puerto Rico, and the Philippines.[175] Five days after Roosevelt had given this speech, James published an open letter in the *Springfield Republican* attacking Roosevelt's rhetoric. He pointed out how Roosevelt's celebration of the strenuous life had obscured the human cost of war "from the sole point of view of the organic excitement and difficulty." He went on to say that Roosevelt "gushes over war as the ideal condition of human society, for the manly strenuousness which it involves, and treats peace as a condition of blubberlike and swollen ignobility, fit only for huckstering weaklings, dwelling in gray twilight and heedless of the higher life. Not a word of the cause,—one foe is as good as another, for aught he tells us; not a word of the conditions of success."[176] James lambasted Roosevelt for turning energy into a dangerous metaphysics. Not only was Roosevelt's "moral fibre" "too irredeemably coarse" to differentiate between energy and its direction, but he ignored the fact that energy had no social worth unless it was directed toward an end.[177] As Sergio Franzese has demonstrated, to counter Roosevelt's emphasis on brute force, James had his "moral equivalent" in mind. With this, he could maintain that the value of energy depended on its moral qualification as organized through actions, ideals, and habits, rather than grounded in a metaphysics that elevated quantity over quality.[178]

It is important to note that, contrary to James's criticism, Roosevelt was sensitive to the importance of control when it came to energy. As Roosevelt dedicated his political career to expanding the armature of government, coming to its apogee in the "New Nationalism" platform of his runaway Bull Moose Party in the 1912 presidential election, he remained one of the most vocal advocates of the "efficiency movement" that promoted the regulation of energy. This went hand in hand with his conservation program, as both stigmatized the sin of waste, especially as fears grew over the rapid depletion of natural resources between 1906 and 1910. Concerned by the threat captured in George Perkins Marsh's book *Man and Nature* (1864), about the destruction of the nation's natural resources, Roosevelt had called for the careful regulation of the nation's resource pool. In 1908

Roosevelt had organized a historic conference at the White House to discuss the relevance of the laws of thermodynamics to all aspects of modern statesmanship.[179] The conference included a broad range of perspectives, from Andrew Carnegie, who discussed the preservation of mineral resources, to Samuel Gompers, the unionist and founder of the American Federation of Labor who stressed the importance of conserving the energies of the labor force, and R. O. Richards, who proposed the establishment of a "national education commission" to organize the nation's "first natural resource"—children.[180] Therefore, just as he celebrated war as the great liberator of energies, he similarly elevated "iron-self-discipline" to control and conserve the more pernicious and unpredictable energies of man.[181] In line with this, Roosevelt popularized the works of the psychological and social theorists G. Stanley Hall and Edward A. Ross, who—concerned with the unconscious "impulses" lurking within the human psyche—proposed a variety of environmental, moral, and governmental controls to discipline and unify the nation's energies.[182] In Ross's mantra there was "always some loose screw or other [that] permits social energy to be wasted, and always some one sees this and wants to tighten that screw."[183] The dual emphasis on release and discipline permeated Roosevelt's political agenda, as he popularized a range of methods that conjured up an energetic surplus, while calling for measures to sublimate and control this social energy—or, in Ross's words, to "tighten the screw."[184]

Riddled with its own internal contradictions, then, the rhetoric that clung to the "strenuous life" placed the onus on the will to harness energy. It transferred the ethical impetus to Ostwald's unhealthy state of "individuation" to conjure up an image of heterogeneous centers generating energy at will. This would reveal itself in different models. Although Osborn projected a perfecting teleological energy, James, the philosopher of pluralism, conjured up a vision of autonomous centers generating energy, harnessing it toward undetermined ends. In contrast, Brooks Adams posited a unifying will, a Washington-type figure that would concentrate the nation's energy. Finally, Roosevelt, the Rough Rider, would both generate and discipline the nation's energy and thereby unite a plural society working harmoniously (in Osborn's "energy conception") toward the gradual unfolding of the nation.

Conclusion

That these alternative energy models gave way to opposing political worldviews would become all too apparent in 1915, after Wilhelm Ostwald was reported in the *Outlook* to have announced that Germany's objective in the current war was to bring the rest of world (living in the base state of "individualism") to the highest level of civilization, epitomized by Germany's "organized co-operation."[185] Germany, he argued, was following the laws of energy, and as a result, it was destined to uplift the rest of the world from the lower cultural state of individuality toward a higher monism embodied by the "energetic imperative." Ostwald's warning, of course, tapped into a wider cultural ambivalence. Before the war, Germany had been celebrated as a paragon of efficiency. With Germany's efficiency now directed toward warfare, the line between efficiency and machine-like inhumanity had been blurred.[186] Increasingly hailed in the American press as the prophet of Germany (having replaced Nietzsche), Ostwald told of Germany's plans to "organize Europe" and then subsequently to expand farther out into the world. From Ostwald's perspective, Germany was seeking to create a "confederation of states" founded on the ethical principle of monism: a "new course for realizing her idea of co-operative energy or social efficiency."[187]

Although Ostwald's "energetic imperative" was at the extreme end of the spectrum, causing a public outcry within the American press (and forcing Leipzig University to provide a public apology), progressive rhetoric struggled to balance both models of energy, preserving freedom while adhering to energy laws.[188] Roosevelt's advocacy of both the efficiency movement and the "strenuous life," for example, shows how these contradictory models had been carefully assimilated within progressive rhetoric. In 1913 Brooks Adams described how this conflict had challenged progressive thought, as he diagnosed the Progressive Party as being "a disintegrating rather than a constructive energy."[189] As the progressive policies implemented by Roosevelt unraveled during William Howard Taft's presidency, Brooks Adams even confessed to Henry Adams that he was "growing doubtful" that the "present disposition" would survive until the year 1930, as he had once hoped.[190] Writing a year after Roosevelt's Bull Moose Party had lost to Woodrow Wilson in the 1912 election, Brooks Adams was sad to declare that Roosevelt "found the pressure towards disintegration resistless."[191] Just as he reached a position to consolidate the "centralization of power," he had been distracted by a coterie of "philanthropists and women."[192]

Indeed, the "New Freedom" campaign that led Woodrow Wilson to the presidency came with the subheading *A Call for the Emancipation of the Generous Energies of a People* (1913), when packaged as a book on the eve of his inauguration. Wilson's "New Freedom" promised to "set the energy and the initiative of this great people absolutely free."[193] It promoted a metaphorical energy that would rise from below, from the myriad centers of the democratic individual: "up from that soil, up from that silent bosom of the earth, rise the currents of life and energy."[194] This was the chaotic energy that would have persuaded Brooks Adams that civilization had sunk into the "chaos of democratic mediocrity" predicted earlier by his brother: "where waters which have fallen to sea levels are engulfed, and can do no more useful work."[195] It was this stance, along with his dislike of Wilson, that drove Brooks Adams to oppose the First World War, deeming it to be an economic war lacking in all ideals apart from "Greed."[196] Furthermore, Brooks Adams identified in Germany a government of great efficiency; as such, rather than fighting for a "doomed" system grounded in "disintegrating forces," he felt that the United States ought to match Germany in the control it had over its energy.[197] In fact, it was Roosevelt's unwillingness to give up on the democratic will of the individual, Brooks Adams concluded, that had been his downfall, resulting in the dissipation rather than the concentration of energy. "Poor Roosevelt," he would later tell Henry Cabot Lodge, "he never could be reconciled. He went on to the end actually believing in 'democracy.' That was what gave him strength. He actually believed in the 'common people.'"[198]

On the eve of World War I, therefore, energy had become central to a much broader cultural conversation about the role of the individual will in the industrial age. For mechanists like Loeb, energy was an ethical principle able to eliminate metaphysical categories, like the will, from the social order. It enabled a monist vision of society in which individualism was subsumed into an organic whole. For others, like William James, the onus was placed on the will to harness energy, providing a vision of autonomous centers self-generating energy. These positions were never exclusive or stable but remained intermingled within the political rhetoric of the Progressive Era. Theodore Roosevelt, for example, combined a variety of different forms of energy within his political agenda, drawing elements from mechanistic science while popularizing a form of vitalist energy that could emerge from the will. Ostwald similarly sought to expand his energetic imperative across the social world, even though he inadvertently reinscribed the will as a controlling device.

And yet, just as Loeb's mechanistic conception of energy was set against Roosevelt's bullish emphasis on the will, this worldview shared many properties with the philosophy of force it sought to counter. By transforming energy into an ethical principle, it converted energy back into metaphysics. Commentators on both sides of the debate thus turned to energy as a moral standard only to confront (consciously or not) the problem that energy only had social value when it performed cultural work. This did not just run in one direction. While Loeb and Ostwald sought to ground social forms in the logic of energy, James externalized it, conjuring energy up as a substance, an aqueous fluid that had to be channeled like water through human ideals and culture. In this model, even though society would mold energy, it remained external to social structures, as human institutions became the container into which energy was poured. Despite the many contradictions implicit within these positions, they pointed toward a paradox that continued to define understandings of energy in the coming decades. As we will see in the next chapter, just as energy was positioned as an external measure for society—privileged for being outside of it—problems remained: How was energy transformed into form? Where did its moral value lie? What was its social measure?

1920–1943

CHAPTER TWO

Energizing Americanness
Climatic Energy and Race

In 1926, a paper by the Russian astronomer and biophysicist Alexander Tchijevsky, read aloud at the American Association for the Advancement of Science in Philadelphia, mapped human excitability against an external quantity of energy from the sun. The sun, he pointed out, underwent rhythmic fluctuations of activity, registered by the appearance and disappearance of sunspots on its surface roughly every 11.2 years.[1] Drawing on a long tradition of racial geography that compared "zones of highest and lowest races and culture" to climatic factors, Tchijevsky asked, "If the different amount of the sun's energy received by different climatic zones has such a great influence on humanity, the question arises: do the periodical changes of the sun's activity resulting in the amount of the emitted innumerable streams of electrical particles and electro-magnetic waves also have an influence upon humanity?"[2] His talk caused a stir, with headlines in the national press reporting his findings: "Mass Excitement Laid to Sunspots," "Man Slave of Sun Spots," and "Sunspots Cause of Wars?," they read.[3] The talk caused alarm not only because "this Russian savant" had suggested that 1927–1929 would be the worst period for radio reception since its inception—given that sunspots, now at the peak of the cycle, were causing considerable interference over the radio—but also because he predicted that this occurrence did not bode well for the fate of man. During this period, not only would the sunspot cycle reach its maximum but also two other cycles would coincide: the thirty-five-year Bruckner cycle and Young's sixty-year cycle. Because of this alarming confluence, Tchijevsky foretold "a great human activity of the highest historical importance which may again change the political chart of the world as was the result of a similar maximum in 1870."[4] Although skepticism was aired over the plausibility of Tchijevsky's thesis (many concluding that it was astrological nonsense), the common disruption over the radio and the telephone was proof that some interfering force must be at work.[5] This led to a spate of hypotheses about how sunspots might be controlling all aspects of life—from cycles of boom and bust, to the production of a fine bottle of vintage wine, to even shortening the hemline of women's skirts.[6] Many began to question what effects the energy from the sun was having on American society.

Interest in the sunspot cycle increased during the 1920s and 1930s, alongside a broader debate about the relationship between the climate, "human energy," and the fate of American civilization. Geographers, ecologists, physicians, astronomers, and popular commentators debated whether human behavior was a register of an energy latent in the environment. Not only did this lead to speculation about the impact of short- and long-term cycles on American society over time, but the climatic extremes experienced across the United States also led to a range of theories about the spatial distribution of human energy across the nation and through different racial groups.

This chapter will examine how *climatic energy* became a framework through which to define Americanness during the interwar period. A virulent strand of "nativism" swept through American society in the 1920s, bolstered by paranoia about immigration and fears over the decline of the so-called Anglo-Saxon race, supported by the pseudoscience of eugenics.[7] Climatic energy, this chapter will show, became a way to define whiteness as distinctly American, at a time when fears about the decline of the Anglo-Saxon race and the nation were widespread in American society.

Scholars have documented the influence that climatic determinists, such as the geographer Ellsworth Huntington, had in cementing the connection between the environment and race.[8] This was rooted in what James Fleming has called a "meteorological Taylorism," which fused the language of industrial efficiency, optimization, and productivity with a concern over the effects of the climate on the human body.[9] Less attention has been paid, however, to the ways in which a discourse around climatic energy was also informed by a wider debate about energy's moral value. Whether or not Americanness could be measured against an energy from the environment raised a basic question about energy's relationship to value, as well as how to control and regulate this climatic energy within the social body.

This chapter will show how a discourse around climatic energy became divided between two contrary positions: one that measured the American body against an external energy in the environment and another that valued "Americanness" for its ability to control and manage the abundant energy supposedly generated by the climate present in a narrow geographic band across the northeastern belt of the United States. It will start by tracing how a discourse on climatic energy, emerging out of the scientific study of sunspots, raised speculation about the impact that the sun was having on all forms of terrestrial life. It will then consider how this research fed into debates around race and Americanness, as energy provided a supposedly "objective" measure through which to define and categorize the human body. In the hands of climatic determinists, such as Ellsworth Huntington, climatic energy was part of a wider discourse about the preservation of the white Anglo-Saxon body, tied up with the language of eugenics, as well as feeding into social debates about immigration and crime. It also informed early discussions about climate change, as growing concerns about a warming climate led to fears that American society (and the American body) was in a state of gradual decline.

Climatic energy was far from an objective standard, however. Rather, it was subject to the same internal paradoxes that surrounded the discourse on "progressive energy." The attraction to climatic energy as a measure of the human body was rooted in a similar desire to privilege energy as an objective measure for society during the Progressive Era. Critics, once again, were quick to point out that energy was not an adequate measure for Americanness on its own. In fact, "misdirected energy" was viewed to be an equally large threat to the social body as the draining of energy due to a warming climate. Prominent reformers had long advocated for the construction of an infrastructure (both physical and moral) to discipline society's energy toward worthy pursuits. This management of energy implied a level of social control, as it was racialized groups (such as immigrants) whose energy, it was viewed, needed to be carefully regulated. From this perspective, Americanness was measured by its ability to channel, regulate, and assimilate the wayward energies of an increasingly fragmented nation. Once again, therefore, the question as to whether energy had an intrinsic social value or whether its value rested on its capacity to be harnessed into social good shaped the debates about energy (and its relationship to race) during the interwar period.

The Science of Sunspots

The suggestion that sunspots could influence terrestrial phenomena was not a new idea in 1926, when Tchijevsky presented his theory at the American Association for the Advancement of Science. As early as 1875, the British economist William Stanley Jevons had developed a theory (outlined earlier by William Herschel) that there was an intimate connection between sunspots, weather, and crops, which produced definite cycles in business activity.[10] It was not until 1908, however, that the solar astronomer George Ellery Hale, founder and director of the Mount Wilson Observatory in Pasadena, argued that the connection between sunspots and other contingent factors was more than just inference.[11] Installing at Mount Wilson a large spectroheliograph that isolated certain wavelengths emitted by the sun, Hale recognized that these mysterious dark spots were solar cyclones, out of which hot gases were whirling around the sunspot centers, emanating highly ionized and electrically charged particles.[12] As these particles entered the ionosphere, the interference caused over the radio materialized the phenomenon, bringing it directly into the living room.

At the same time as radio disturbance demonstrated that sunspots were having a tangible influence on earth, the astronomer Andrew Ellicott Douglass, from the University of Arizona, discovered a long-range historical record of their impact in the rings of sequoia trees.[13] Having recognized the existence of sensitive rings that swelled with increased precipitation, Douglass identified a variation occurring roughly every eleven years. By cross-referencing records, tree rings made visible the same climatic variable working in geographically diverse microenvironments.[14] This provided a long-range archive that aligned diverse localities in historical time with changing climatic conditions caused by variations in the solar constant.[15]

This was supported by research into animal cycles. As early as 1924, the British animal ecologist Charles Elton published research on the synchronicity in the fluctuation of lemming populations across the arctic regions and mountains of southern Scandinavia, connecting this rhythmic fluctuation to the sunspot cycle.[16] As interest in the subject grew, in 1931 a high-profile conference would be organized to study the impact that these cycles were having on natural life. The Matamek Conference on Biological Cycles, organized by industrialist D. C. Copley Amory, brought together a prestigious coterie, which included Elton, Ellsworth Huntington, and the ecologist Aldo Leopold, to study the economic, sociological, and agricultural benefit of decrypting these "powerful natural forces." During the conference, many different types of cycles were discussed, from a four-year cycle in snowy owls, described by A. O. Gross, to a ten-year cycle found in nonmigratory birds, rabbits, and their predators, as elucidated by William Rowan.[17] Even though the cause of these cycles remained unclear, it was concluded that all sorts of cyclic phenomena "must be controlled, though not necessarily caused, by some outside forces which dominate all forms of life."[18] It was not just animal populations that were subject to these outward forces; papers also made a direct connection to human populations.[19]

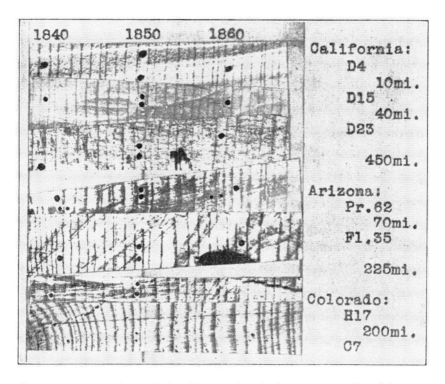

Illustration mapping the similarity of the tree rings in 1851 across 750 miles of the country. A. E. Douglass, "Evidence of Climatic Effects in the Annual Rings of Trees," *Ecology* 1, no. 1 (January 1920): 26.

Alongside the growing interest in the effect of solar cycles on plant and animal populations, disruption on the radio, caused by the height of the sunspot cycle in 1927, materialized the phenomenon and brought it into the living room. In 1929 the popular science writer James Stokley pondered in the *Science News-Letter* whether "the present popularity of radio is largely due to the fact that there were few sun spots in 1923!"[20] Recognizing that "in October of this year radio reception will probably be poorer than it has been since broadcasting came into popularity," he predicted that "not until the year 1934 may we expect really good radio reception, comparable with that which we enjoyed in 1923!"[21] While for many this disruption over the radio was a nuisance, to the astronomer and radio enthusiast Harlan True Stetson it proved a rare opportunity to establish a new discipline: "cosmecology."[22] In his book *Earth, Radio and the Stars* (1934) Stetson outlined how, with the aid of the radio, cosmecology would measure the relation of terrestrial earth to the cosmic scheme.[23] Drawing on the emerging field of ecology (led by many of the participants at the Matamek Conference), cosmecology was based on the principle that "the mysterious electron, the fundamental building block of all matter, which dances in your radio tubes to the tune of your favorite jazz, is dancing in the remoter stars of the galaxy."[24] Stetson argued that by monitoring radio interference one could measure the disturbance generated by sunspots on all forms of organic matter, including human populations. Radio enthusiasts, like Orestes Caldwell from the Federal Radio Commission (and the editor of *Electronics*), reiterated this argument, noting the correlation between the disruptive effects of sunspots on the radio and the remarkable variation in woodland animals.[25] Caldwell pointed to the obvious correlation: "more sunspots; less rabbits; few sunspots, lots of rabbits."[26] Replacing "radio static" with "rabbits," Caldwell went on to explain how the variation in animal populations was probably due to the same "excess of ultra-violet which upsets our radio in eleven-year cycles."[27]

"An Index of Mass Human Excitability"

Once it became evident that interference over the radio paralleled the disturbance of organic matter, speculation as to how society was being affected by sunspots became rife. Edward S. Martin, editor of *Harper's Magazine*, dedicated two of his "Editor's Easy Chair" reflections to sunspots in the March and November 1927 editions.[28] In the March editorial "Sun Spots and Politics," Martin discussed Tchijevsky's recent talk at the American Association for the Advancement of Science. Recognizing Tchijevsky's essential thesis that "the world does get very crazy at times," he reflected on the recent situation. He pondered whether sunspots were to blame in 1912 for "such a curious quiver of behavior and such an extraordinary craze for dancing."[29] Suggesting that this "craze for dancing" may have been "one of the preliminary periods of the sunspot cycle which introduced the Great War," Martin turned to the more familiar phenomenon of radio "rays" to elucidate this somewhat tenuous thesis.[30] Martin noted, "This Russian informant's sunspot theory—that's rays."[31] In a blithe manner Martin explained that all one need do is imagine that these "rays" are a "substance" like radio rays, or the kind of rays "actors or great orators" are known "to produce" and "shoot out," and then it becomes obvious how they might work on us. It was no longer such a bizarre idea, he pointed out, to imagine how "rays" may operate, since the "idea that force is exerted only by visible and physical means does not go so strong as it used to."[32] Now that "rays" are a common phenomenon, Martin noted, "we are getting more familiar with the idea of powerful and important forces that are imponderable, invisible and very imperfectly understood."[33]

At the 1927 meeting for the American Association for the Advancement of Science, Tchijevsky demonstrated how sunspots directly influenced the excitability of human beings.[34] Tchijevsky mapped a correlation between the eleven-year sunspot cycle and human excitability. Calculating both the quality (the importance) and the quantity (the mass involved) of "carefully chosen" historical events, Tchijevsky constructed an "index of mass human excitability." This cycle progressed through four stages. The first stage involved three years of minimum excitability. The second consisted of two years of increased excitability. This was followed by three years of maximum excitability. The fourth stage represented three years of decline back to a minimum that closed the cycle.[35] During the first stage, there was little unity within the masses and a marked indifference to any idea or action that upset the peace. Evidence of this could be found in autocratic forms of government and a rise in peace treaties.[36] In the second stage, the human masses underwent a process of unification, as national and political parties galvanized around an issue.[37] Separate causes crystalized around one psychic center, for example, a prominent idea, creating a greater integration of the masses, a phenomenon reflected by the tendency toward majority governments.[38] In the period of maximum excitability, the masses got "more impatient, nervous, and exacting." It was at this time that the greatest events in human history occur: wars, insurrections, rebellions, campaigns, migrations, and expeditions.[39] These periods witnessed the rise of leaders, the triumph of ideas by the masses, democratic participation, democratic social reform, and high levels of parliamentary practices.[40]

Each of these stages matched the activity of the sun, with the period of highest excitability aligned with the solar maxima. This reciprocity, Tchijevsky attested, was so accurate that one could even map the influence of the sunspot cycle in the life of individuals, such as Napoleon Bonaparte. He explained: "The maxima and minima of Napoleon's activities correspond to the same periods in the sunspot activity. Thus, 1809–1811, minimum of sunspot activity (Wolf) and no campaigns; 1804 maximum of sunspots and the greatest campaigns of Napoleon with the title of emperor."[41] In the *Harper's* editorial Martin adopted Tchijevsky's thesis but turned to the central issue of the day, the upcoming election, to ask, who would "the sun spots work for, Smith or Coolidge"? The former candidate, Martin deliberated, being the more "revolutionary" figure, "must have good rays shot into him."[42] Suggesting that this candidate's "heightened energy" may appeal to a more energized public, Martin expressed his hope that "the sunspot rays" that the Russian scientist predicted may "get us going again."[43] All that was needed was something "with a kick in it."[44]

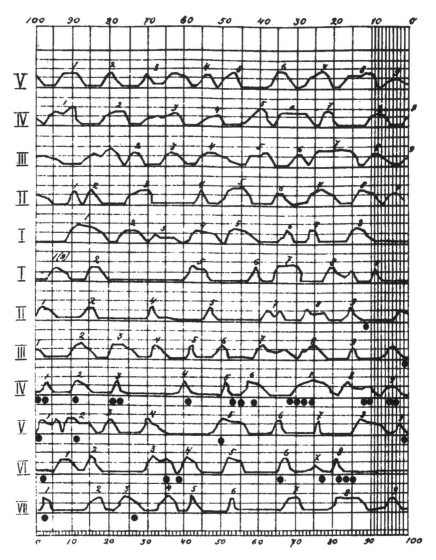

Above and opposite: The fluctuation of the universal historical process across the globe, between the fifth century and the nineteenth century. The lines mark the quantity of historical events, and dots mark the pre-telescopic and later-astronomical data of the sunspot maximum. A. L. Tchijevsky, "Physical Factors of the Historical Process" (1926), *Cycles*, January 1971, 24.

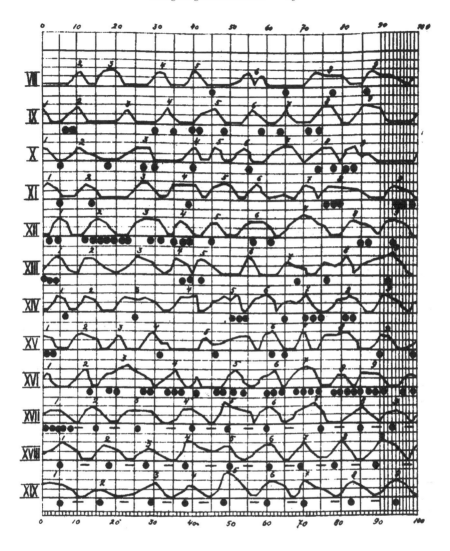

If Martin hoped this excess energy would sway the presidential campaign in favor of Smith, by November he had grown concerned about the spots. In his November editorial, "Sunspots and Justice," he worried that "the unusually agitated condition of the sun presaged corresponding agitations in the minds of men."[45] Whereas in March Martin believed that sunspots might energize the American public, by November he blamed them for the recent social unrest. This time he posited the uproar surrounding "The Hall-Mills case, Lindbergh and the flyers," and "the case of Sacco and Vanzetti" to be "jobs of widespread mental disturbance" caused by sunspots.[46] Where a few months earlier he trusted that increased activity would give the American public a good kick, now he wanted the sunspots to disappear. Martin was not the only concerned party. A few years earlier, anxious as to whether there might be some deeper cause "affecting everyone in the neighborhood at the same time and in the same general way," Forum magazine launched their own investigation to determine how sunspots might be working on their readership.[47] To do this, it asked its readers to record how they felt each morning on a questionnaire, which included the categories "fine," "energetic," "lazy," "nervous," and "rotten."[48]

Alongside this popular speculation, other studies began to explore the extent to which solar cycles were affecting physiological and psychological functioning.[49] The psychologist Raymond Holder Wheeler, for example, explored how the solar constant affected human thought and attitudes. Based at the University of Kansas, Wheeler was an advocate of Gestalt psychology in the United States, with a focus on the impact of climate on world history.[50] In his early book *The Laws of Human Nature* (1931), Wheeler described how "the logical significance of life and will are found in the unity of meteorological, magnetic, temperature and gravitational fields."[51] These, what he would later call "unified-energy-fields," provided a schema to understand all elements of the human world—the will, personality, and culture—all understood as a direct materialization of energy.[52] Because of this continuity, Wheeler was adamant that human action had to be rethought. It had to be expanded to include "any activity whether in the outside world or in ourselves—physical, chemical, biological, psychological, social, political, animate or inanimate. It includes all motion, all flowing, all processing, whether of falling bodies, winds, electrical currents, chemical reactions, or human achievement. It includes the mental and the physical, the material and the immaterial."[53] Following the belief that "human will" and "intellect" operate in a "unified-energy-field," Wheeler maintained that culture not only was the direct materialization of energy but also could be directly aligned with climatic pulsations controlled by the sunspot cycle.

He later developed this theory in an article titled "The Effect of Climate on Human Behavior in History" (1943), where he described how whole epochs could be characterized through their different energetic conditions. For example, Wheeler noted how sunspots changed the world climate, which became warmer at periods of sunspot minima and colder during periods of sunspot maxima.[54] This led to periodic differences that had their own set of unique characteristics—"warm-wet, warm-dry, cold-wet, and cold-dry"—leading to different cultural types.[55] For example, southerners and northerners, Wheeler believed, "have very recognizable personality types" in times of climatic fluctuations: "Northerners take on the psychological characteristics of their more southern neighbors during warm times, and southerners become more like their northern cousins during cold times. During cold times the qualities of the northerner become more 'northernish,' while during warm times those of the warmer climates become more 'southernish.'"[56] Drawing on long-standing racial stereotypes, warm-climate people, he argued, tend to be prone to "indulgent, passionate, and 'sexy' behavior."[57] As a result, during warm periods there was an increase in "fanaticism, cruelty, tyranny and dictators."[58] However, if "Hitler" was a "typical hot-drought specimen," cold periods were more democratic, fostering the characteristics of freedom and individuality. This shift was manifest in culture and trends in art and literature.

Due to this confluence, Wheeler mapped his culture curve against the climatic one found in the tree ring record. Thanks to the "stories" told in tree rings, Wheeler concluded, he could now align events that occurred independently in the eleventh century. Recognizing a period "so warm that trees grew in Greenland," Wheeler set out to align all of "the Hitlers of that day"—"Basilius II of Byzantium,

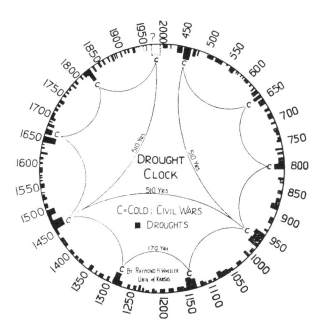

Clock of cold droughts and civil war. R. H. Wheeler, "The Effect of Climate on Human Behavior in History," *Transactions of the Kansas Academy of Science* 46 (April 1943): 38.

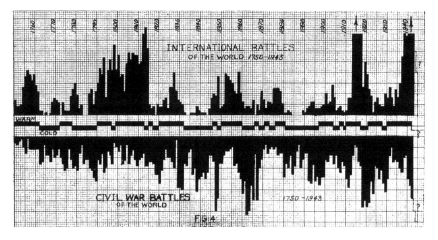

The distribution of international and civil war battles occurring in all known parts of the world from 1750 to 1943. The height of the column is a summation of the activity in a given year. Warm and cold years are indicated in the middle of the graph. R. H. Wheeler, "The Effect of Climate on Human Behavior in History," *Transactions of the Kansas Academy of Science* 46 (1943): 50.

Sviatopolk of Russia, Olaf of Norway, Cnute the Mighty of Denmark"—seeing each of them as a historic record documenting a certain level of energy.[59] Wheeler's "unified-energy-field" was at the extreme end of the spectrum, but as others, such as the economists and industry men gathered at the Matamek Conference, became increasingly aware of the impact climatic cycles were having on terrestrial life, solar cycles provided an objective measure against which one could trace trends and movements within human culture and spirit. They provided a historic archive within which a whole range of human behavior could be mapped against an external measure in the environment.

There was little consensus, however, as to how energy was materialized by these forms. After all, they might bring on outbreaks of crime, but as Martin realized, they could also sway the upcoming election. In addition, there was no clear indicator of the relationship between energy and its secondary effects. Tchijevsky himself recognized this fact, pointing out that "the sun does not oblige us to do this or that; it only obliges us to do *something*."[60] Because of its undetermined nature, Tchijevsky hoped that humans might be able to "find ways to a humanitarian use of the mass upheaval, by means of preliminary propaganda for some undertaking of great public interest and importance which is to be completed in the period of the maximum of excitability."[61] Harnessed away from war, this energy could go toward productive activities: "scientific expeditions, sport competitions, building of stupendous structures (bridges, canals, railroads, whole towns, etc.), collective theatrical art, collective creative art."[62]

Although Tchijevsky—drawing on James's "moral equivalent"—maintained space for human direction in the "universal historical process," the rigid causality of his forecasting model left him, as well as those who posed similar theories, open to criticism. One of the most astute attacks came from the Harvard sociologist Pitirim Sorokin, who explored the many conceptual fallacies embodied in this field of research. In his book *Contemporary Sociological Theories* (1928), Sorokin dedicated an entire chapter to criticizing the so-called Geographic School led by Huntington, while also devoting another chapter to the school of "Social Energetics" fronted by Wilhelm Ostwald.[63] For Sorokin, the premise that sunspots could operate as an "external agent" was just as potent a convention as other "philosophical abstractions, like Destiny, Providence, and so on."[64] Its whole premise depended on reducing the totality of social experience to a commensurable standard. It took entirely incommensurable conditions—"war" and "excitability"—and tried to find a common measuring stick in a quantity of energy. This was a futile maneuver, for it was impossible, he explained, "to compare and co-measure the role of sun-spots and that of fear or lust; the weight of fear and that of the universal law of struggle for existence. . . . It is evident that they cannot be measured, or even roughly appraised in any comparative way; there is no measuring stick applicable to all of these."[65] War, Sorokin pointed out, could not be measured against heightened levels of excitability, for war, being a complex social experience, could not be reduced to a physical magnitude. Similarly, it was impossible for emotions, such as fear or lust, to be reduced to comparable quantities, for, as Sorokin argued, there was no "measuring stick applicable" to such intangible conditions. Not only was it an arbitrary convention to balance incommensurable sets of experiences, but Sorokin also remained adamant that one could not separate a single set of relations from the infinite other influences that operated on society at any one time. It was tantamount, he maintained, to seeing only one set of cause and effect. Because of this fallacy, he pointed toward the error of causality implicit in this understanding of energy: if one could go as far as to assume a direct physical magnitude between two kinds of energy ("climatic" and "spiritual"), this did not account for why energy was harnessed into a particular form. To clarify this, Sorokin asked the reader to "take the behavior of individuals, A, B, C, D":

> Can we explain the immense variety of their actions through the principle of physical mechanics, through that of inertia, gravitation, or by means of the principles of levers of the first and of the second orders, and so on? Do they help us to understand

why A becomes a hermit, B marries, C dies on a battlefield, D writes a poem, and so on? Do these principles throw a light on the religious, political, aesthetical, and other social phenomena? Can they explain why the history of one people has developed in one way, and that of another in quite a different manner?[66]

Contrary, therefore, to the widespread belief that sunspots could operate as a suitable predictive tool through which to manage society, Sorokin had to dispute the *"illusion that there are definite natural laws and an iron determinism which resides in the social processes, and that a knowledge of these laws is possible."* Pointing out that the advantage of sunspots was to provide the illusion of a forecasting mechanism, Sorokin declared that this fallacy rested on an inaccurate notion that the development of society could be "accurately predicted." In the wake of the "99,999,999" societal variables that could not, sunspots gave the illusion of complete knowledge.[67] Energy was a tidy convention that united an unintegrated culture through a coordinating nexus; it allowed one to think in universal causes.

Ellsworth Huntington's Energy Determinism

Despite the reproofs that came from critics like Sorokin, it proved easy to overlook the inconsistencies that emerged when energy was applied as a continuous measure across social experience. Simplicity, for those who applied this model, provided a much stronger explanatory tool than complexity. In the hands of the climatic determinists, energy became a useful tool through which society could be compared against an external measure found in the environment. The area in which this gained most traction was in the study of civilization. The association between climate and civilization sat within a long tradition of "climatic physiology" originating from Hippocrates's *On Airs, Waters and Places*, in which human diversity was accredited to the surrounding environment.[68] This had taken on a new life in the hands of climatic determinists, such as Ellsworth Huntington, who looked to energy as a quantitative measure to assess human progress.

Denied a tenure post at Yale, and largely derided by the geographic community, Huntington wrote for the popular press, becoming a familiar voice across news and magazine outlets—an authoritative guardian watching over the fecundity of America's energy.[69] He thus joined the ranks of other prophets, such as Brooks Adams, who were preoccupied with the relationship between energy and civilization. And yet, whereas Brooks Adams had focused his attention on the control of energy, Huntington gathered large bodies of empirical evidence that proved civilization to be a physical register of its climatic energy.

Since his earliest book, *The Pulse of Asia* (1907), Huntington had posed a range of hypotheses as to how climatic energy was translated into equivalent levels of physical, mental, spiritual, and creative energy that built a civilization.[70] Huntington, however, elaborated this argument most fully in his book *Civilization and Climate* (1915), where he mapped climatic energy against "levels of civilization."[71] He did this by approaching an international coterie of historians, diplomats, colonial officials, travelers, educators, and businessmen, asking them to rank where they judged the most civilized societies to be. To gauge the appropriate measure of civilization, Huntington gave a range of qualitative standards against which this should be judged. This included "the power of initiative, the capacity for formulating new ideas and for carrying them into effect, the power of self-control, high standards of honesty and morality, the power to lead and to control other races, the capacity for disseminating ideas, and other similar qualities which will readily suggest themselves."[72] How merits such as "power of initiative" were to be read as a marker of human energy was not explained, although Huntington implied that evidence of high levels of energy could be found in ideals, respect for law, and the capacity to carry out difficult enterprises.[73] However, not only was it unclear how credentials such as the "powers of self-control" stood as a record of human energy, but his description of climatic energy was also similarly confused. The climate, after all, is not synonymous with energy, and in order to understand it as such, Huntington had to collapse different factors like humidity, temperature, and variability into a common standard. To cement the image of climate as an energy force, Huntington drew on potent metaphors, describing it as a "driver upon his horse" and the "steam" that fuels a train.[74]

Huntington reiterated this comparative action countless times (in 28 books and over 240 articles), conjuring up ever more elaborate scenarios of how climatic energy was converted into human energy. Some of these pretended to be empirical, such as the studies he completed as chair for the Committee on the Atmosphere of Man, which, as part of the National Research Council, set out to explore climatic influences on the efficiency of factory workers. However, Huntington had a propensity for leaps of the imagination and used energy as a conceptual device to map ever more hyperbolic sets of relations. In *World-Power and Evolution* (1919), for instance, he speculated as to how the storm belt that occurred over Italy from 1325 to 1460 provided a large excess of climatic energy that contributed to unique levels of mental stimulation.[75] The fact that this climatic event coincided with the "Revival of Learning" during the Renaissance proved "that during the fourteenth century the mental activity of Italy was higher than at any time since the days of Rome."[76] Although the Renaissance did not happen until a century later, Leonardo da Vinci and Michelangelo, Huntington pointed out, could not have reached the heights they did without the aid of men such as

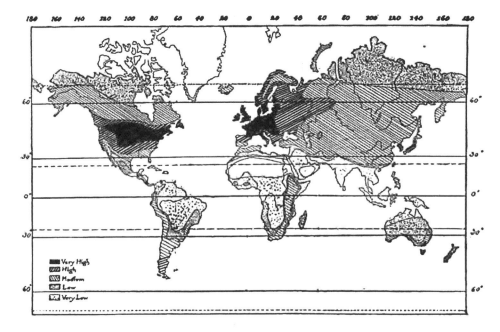

Map charting levels of human energy against climate. E. Huntington, *Climate and Civilization* (New Haven, CT: Yale University Press, 1915), 142.

Giovanni, Pisano, and Giotto who worked in this invigorating climate. These men were the "original pioneers who did most of the inventive work, while Leonardo and Michelangelo were the reapers."[77] In Huntington's framework, therefore, civilization was not attributed to creativity, artistic prowess, or culture but was purely a record of the amount of human energy as it evolved through time.[78]

Huntington's map of civilization drew heavily on, and reinforced, racial tropes circulating at the time. After all, Huntington's map of civilization (compiled, as it was, mostly by Europeans and Americans) confirmed what most of his readers already assumed: the northeastern band of North America (along with Western Europe) was endowed with the highest levels of human energy and as a result was the cradle of civilization. Levels of civilization gradually declined the farther one moved away from "the temperate zone" on account of the lower levels of human energy derived from the climate. Huntington thus affirmed what climatic physiologists had long argued: that racial difference was a result of climatic variation and that the superiority of the white Anglo-Saxon race (over other racial groups) could be attributed to the potency of climatic energy that had driven the evolutionary process forward in those regions. Huntington's map thus used human energy to place white Anglo-Saxon bodies at the top of the racial hierarchy.

Despite its flawed logic, Huntington's map of human energy fed into contemporary discourses on race and the body, informing debates about colonial expansion and immigration. The superiority of the "temperate belt" over the stultifying tropical regions, for example, as Daniel Bender has shown, justified the colonial expansion of the United States into the Philippines, while also offering a powerful warning against annexation and the prospect of white settlers in these regions, who would suffer in the tropical climates.[79] This rationale depended on long-held colonial stereotypes about how the heat of the tropics bred laziness, hypersexuality, drunkenness, insanity, crime, and suicide, not only in the native born but also in the "white man" who succumbed to this environment. The same theories could also be applied at home to justify, as Huntington would, the difference between the industrial North and the underdeveloped South, where the extreme heat of the region had led to its "backwardness" and the racial degeneration of both African Americans and the whites who settled there.[80] Even African Americans who had migrated to the temperate zone of the Northeast, in Huntington's scheme, were still defined by the tropical climate of Africa, unable to escape the stultifying effects of the heat on the long-term evolution of the race.[81] Characterized by the same tropes of laziness, indolence, and disobedience as those in tropical regions, African Americans were perceived to lack the energy and vigor to participate fully in mainstream American life. Therefore, human energy, as interpreted through Huntington's climatic theory, provided one of the strongest markers for race. The white body was characterized by high levels of energy, vigor, and efficiency. In contrast, the black body was defined by torpor, friction, and lack of momentum.

Huntington's interest in long- and short-term climate cycles, moreover, extended his theories about climate and race back through time. He described how climatic centers had shifted throughout history, drawing with them the tide of civilization. He emphasized this thesis in his subsequent books *Climate Changes: Their Nature and Causes* (1922) and *Earth and Sun: A Hypothesis of Weather and Sunspots* (1923), where he outlined how changes in the solar constant had driven civilization's progress. To explain this, he drew on the work of the sociologist S. Colum Gilfillan and his theory of "The Coldward Course of Progress" (1920). Gilfillan, who was heavily indebted to Huntington, argued that civilization followed a northward progress toward colder climates—a dual result of the technological capacity to transform inhospitable environments and the mental and physical stimulation of colder regions (which give a "population energy, achievement, and empire").[82] Tracing the southward direction taken by the Roman Empire during its collapse (as its center shifted from Rome to Carthage),

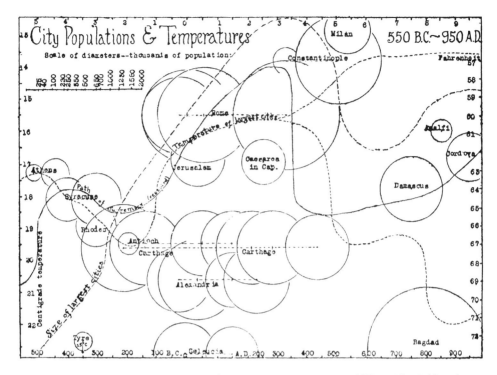

City populations and growth mapped against temperature. S. C. Gilfillan, "The Coldward Course of Progress," *Political Science Quarterly* 35, no. 3 (1920): 393.

Gilfillan concluded that the move toward warmer climates was synonymous with civilization's collapse. The "coldward course of progress" was ongoing, and consequently Gilfillan predicted that New York—now the cradle of civilization—would soon be replaced by Detroit, the heart of the automobile industry.[83]

While Gilfillan explained this advance spatially—as a movement northward—Huntington emphasized the temporal nature of this process. By doing so, Huntington cemented the notion that progress was fragile, dependent on a changeable climate that controlled a civilization's energy and vigor. Were the climate to gradually warm up, North America, as the cradle of civilization, would follow the course of the Roman Empire and undergo a political, racial, and mental decline.[84] Huntington made this connection explicit, pointing out how within his theory of "climatic pulsation" a "most disquieting vision of retrogression is disclosed."[85] He thus concluded *Civilization and Climate* with apocalyptic flair, conjuring up a new Dark Ages that would come about in a few hundred years, once the center of energy had been relocated from the United States to "Egypt, Mesopotamia, and Guatemala."[86] As new and old world powers swapped positions, "races of low mental caliber," he worried, "may be stimulated to most pernicious activity, while those of high capacity may not have energy to withstand their more barbarous neighbors."[87]

Huntington's theory of climate and civilization played into long-held fears over the limitations of Western civilization, its cultural values, and its social hegemony, as it set itself in opposition to the tropical regions, which came to represent a sphere in which Western man and his cultural values would perish.[88] As such, his theory reflected common racial anxieties about the decline of national character and degeneration that had long saturated the popular consciousness. Huntington's drive to conserve the nation's "racial stock" would eventually lead to his appointment as president of the American Eugenics Society in 1934, although long before that his books had articulated familiar anxieties over "racial degeneration."[89] During the twenties, he had studied the problem of immigration from this perspective and frequently testified before the Immigration Committee in Washington, DC, about its negative effects on the United States.[90] The racial characteristics of immigrants, he believed, had evolved in regions with less optimal climates and therefore, he argued, would not be able to meet the high-energy demands required in industrial America.[91] His books and articles were riddled with warnings about the fragility of the native-born American, especially those of Anglo-Saxon or Nordic stock, who he believed were under attack from a "floodtide" of immigration.

Huntington's theories on climate, therefore, defined Americanness in energetic terms. The variability of the American climate, Huntington would often point out, contributed to America's exceptionalism: it contributes to our "intense activity," enabling us to excel in "material wealth" and "our passion for 'putting things over.'"[92] There were, however, challenges that came with such abundant energy: "it does not make us intellectual, or thorough, or tactful."[93] However, with the right checks on energy, Huntington believed that "we can perhaps learn to regulate our powers of speed more reasonably, and thereby improve our character."[94] The state of American civilization was, therefore, balanced on a precipice: threatened by too much energy, the distinctive climate had bred that peculiarly American "nervousness" recognized by George Miller Beard. At the same time, as the American climate suffered from intensely hot and humid summers, and below the Mason-Dixon line temperatures reached scorching, the nation was never far from the "tropical" conditions that spelled its collapse.

The Problem of "Misdirected Energy"

Both the quantity of energy and its control thus emerged as mutually dependent problems. A strong energy drive was paramount for civilization, but without appropriate checks this energy would lead to its downfall. The Cincinnati doctor Clarence Alonzo Mills did the most to popularize the latter point of view. Mills was the author of a series of popular books, including *Living with the Weather* (1934), *Medical Climatology* (1939), and *The Climate Makes the Man* (1942), as well as countless articles published in the popular press. He had been stationed in Panama and spoke often of experiencing the depressing effects of tropical heat. In *Living with the Weather*, Mills described the potency of North America's "energy stream." He painted a picture of Mrs. R., "stretched out on the couch, attempting to relax and quiet her frayed nerves."[95] The children were finally in bed, and all was calm. In distress, Mrs. R. exclaimed, "Why must they be so trying when they are awake and yet so angelic when they are asleep that they shame me for my impatience and irritation?"[96] Noting that Mrs. R. was not alone, that her frustration with her children was a particularly American problem, Mills consoled, "Only in northern states do we see babies that begin to bounce almost as soon as they are born, growing into children never still a waking minute, and later developing into young people wanting to be on the go every night."[97] The speed that drove American youth, Mills explained, was due to the body "whose energy level . . . may be thought of as very much like the available horse power of a motor."[98] Due to the potency of this climatic drive, energy in North America was being spent in youth at a rapid rate. Taking this energy to be a limited resource, Mills feared that most of it was being wasted in "flaming youth," leaving little energy for the more mature citizens to channel toward higher ends.[99] Because of the potency of this drive, only a few rare citizens, Mills recognized, have any vitality left over the age of fifty, which thereafter can be applied toward the building up of civilization.[100] To counter this threat, Mills put forward an argument for the unique responsibility that the nation had to channel and control its abundant energy.

By imparting this warning, Mills drew attention to the fact that although climatic energy provided a good measure of civilization, this energy had to be carefully controlled. In doing so, Mills drew on a long tradition of "positive *environmentalism*," a guiding principle of the reform movement that had emerged during the Progressive Era.[101] Reformers believed that the key to solving the problems of industrial society, such as crime, intemperance, prostitution, and immorality, was by improving the physical and moral environment. This was often framed in energetic terms, with many of society's social ills attributed to "misdirected energy." For example, in *The Spirit of Youth and the City Streets* (1909) Jane Addams, founder of Hull House, had concluded that "young people are overborne by their own undirected and misguided energies."[102] The badly designed city, along with the rise of monotonous industrialized labor, Addams believed, had blocked the channels through which the vibrancy of youth and "the wine of life might flow," leading to an explosion of crime and juvenile delinquency.[103] This idea persisted, and by 1928 the Baumes Committee, set up to investigate rates of crime and juvenile delinquency in New York, had concluded that two-thirds of this activity could be attributed to "misdirected energies."[104] The committee suggested that by adjusting juveniles' moral and material environment "the gang energy and spirit of adventure" could be harnessed "in behalf of something worthy—something that challenges their imagination, their courage, their ingenuity, their strength, their skill, their loyalty to one another—the moral equivalent of thievery and other adventurous delinquency."[105] From this perspective, the role of the reformer was to construct the physical and moral channels through which to control this energy.

Nowhere was this view more prevalent than in the playground movement, which advocated a network of urban parks in tenement districts as a remedy for juvenile crime and delinquency. The playground movement reflected a growing concern about how to control the energies of youth, which could not find an outlet in the overcrowded tenement districts of urban America. Urban playgrounds were promoted as an important "escape valve for pent-up energy."[106] Harriet Heller, for example, who had spent three years studying the problem of juvenile crime, believed that burglary, arson, street fighting, theft, and gang warfare committed by bored youth could be prevented by equipping the city with more playgrounds, which "would eliminate most of the misdemeanors arising from misdirected energy."[107] The playground was, in her words, a "social dynamo gathering invisible, wasting force and transmuting it into vital, electrifying power that shall go thrilling and throbbing to the remotest corners of the community, stirring to the achievement of better homes, better children, fuller and better lives, higher and more joyous living."[108] The idea that social ills could be attributed to misdirected energy stood the test of time, and as late as 1938 Louis B. Nichols, aide to head of the Federal Bureau of Investigation J. Edgar Hoover, argued that "misdirected energy" was more important than poverty and illiteracy in driving crime.[109] The most notorious criminals came not from homes with little material comforts but from ones that lacked "proper spiritual training . . . , with indifference on the part of parents in directing the energies of children in a worthwhile manner."[110] To avoid a life of crime, the nation's youth, he urged, should "direct their energies toward principles represented by their association," such as creative pursuits and wholesome forms of entertainment.[111]

The need to design a strong physical and moral infrastructure to channel energy was repeated by figures like Mills, who stressed the importance of controlling the excess energy stimulated by the vigorous American climate. For example, in an article published in *Harper's Magazine* in 1941 he refuted the cultural stereotype that women in the tropics reached sexual maturity earlier than those in temperate regions.[112] In fact, his own findings had proved that women in Ohio and similar latitudes grow "most lustily" in temperate climes.[113] Rather, it was the moral constraints placed on energy in colder regions that made its inhabitants more sexually reserved. Mills believed he had proof of this fact. He reflected that, "after twenty years spent in trying to direct intelligently the energies of three growing youngsters in Cincinnati, I often find myself envying tropical and oriental parents their more docile, better-behaved offspring."[114] Consequently, Mills suggested, "temperate zone-races could far out-breed those of the tropics if they consorted with the same lack of restraint."[115] Because of the added pressure placed on the American youth by this strong climatic drive, Mills believed that parents and educators had a collective responsibility to construct the physical and symbolic infrastructures to regulate energy. The "'thou shalt nots' of Christianity," he added, were imperative for the "energetic people of cool climates where vitality runs high and submission to every impulse would bring chaos."[116] Similarly, in cold regions, Mills strongly advised against popular activities such as competitive sports, "theatre, both stage and screen," "jazz music," and "exciting mystery stories in books or over the radio."[117] Although these activities provided a "safety valve for the escape of excess energy," they "too often rouse emotional storms and excite the nervous system under conditions where physical outlets of energy are suppressed."[118] In order to harness the nation's energy, Mills prescribed two possible solutions. The first was that the physical and moral environment should be constructed to divert this "lusty energy" toward productive ends. Light exercise could be encouraged through the development of an infrastructure of walking routes through the country. The second was to manage the quantity of energy by utilizing technologies to control the indoor climate, through the use of heating devices and air conditioning.[119]

These youthful energies were not the only threat the nation faced. As concern over immigration fed into broader fears that the nation was being flooded by outsiders with dubious moral habits, immigrants were another demographic whose energy required careful observation. The immigration acts of 1917, 1918, 1921, and 1924, which curtailed immigration by 85 percent and stigmatized

Southern Europeans, Eastern Europeans, and East Asians, would be justified along these lines.[120] Mills claimed that immigrants from Northern Europe (Britain, France, Germany, and Scandinavia) had no problem acclimatizing to the laws and customs in North America. Habituated to equally vigorous climates at home, by the time they settled in the north of the country, they had the cultural and social inhibitions in place to regulate the high levels of vital energy found in the American continent: "past centuries had adapted them to habits of life, social customs and the inhibitions which go with a fairly high degree of bodily energy."[121] In contrast, immigrant groups from warmer regions posed the biggest threat to the United States because they lacked the regulatory edifices and moral conduct to control the levels of energy generated by the climate of North America. Mills remained adamant that their low levels of vigor in the warmer climates did not require "moral and social restraints" to channel energy. However, the sudden influx of energy on arriving in the United States meant that immigrants had no control over their energy, and as a result they had come into conflict with the law, leading to the recent crime wave. Mills conjured up a scenario in which the immigrant's energy, lacking the controls that kept the North American and European in check, was running wild, threatening the stability of the entire social edifice. As a result, Mills was able to diagnose that it was the "new-found energy of the new arrival which causes him to be such a misfit in his surroundings, affecting not only himself but others about him."[122] Echoing the sentiments of racial hygienists, Mills insisted that unrestricted immigration was unadvisable on account of this uncontrollable energy and should not be allowed to continue.[123]

A Warming Climate and a New Dark Ages

According to this line of thought, the United States was menaced by two contradictory factors: "misdirected energy" was one threat, but equally, due to long-term shifts in the climate, there was no guarantee that the energy revenue provided by the stimulating climate would last in perpetuity. The latter problem, Mills believed, posed one of the biggest challenges to the geopolitical power of the United States, whose growing prominence on the world stage was a result of its fortuitous energy basis. As such, he reiterated the familiar caution that although the climate had driven the progress of American civilization, it could just as easily lead to its collapse. Whereas Huntington took a long view, forecasting this devolution over the next hundred years or so, Mills was more pessimistic, suggesting that this process was already under way. In his subsequent book, *Climate Makes the Man*, he described how the invigorating weather that had made

America great was swinging toward a trend of milder winters and hotter summers, leading, as one reviewer noted, toward a "new Dark Ages."[124] Mills had been reiterating this thesis for some time.[125] In doing so, he added another voice to a body of speculation that recognized that the climate was heating up.[126] During the 1930s, the press frequently carried reports documenting how older people had started to believe that the climate was changing; the vicious winters of the 1890s seemed to be getting milder as summer temperatures rose.[127] Meteorologists confirmed this hearsay, agreeing that a warmer trend was indeed under way, brought about by long-term cycles in the climate.[128] By 1939, scientists were noticing that the annual average temperatures over the decade had been the hottest on record, a fact made evident by the Dust Bowl. Other studies on melting ice in the Arctic, as well as Douglass's work on tree rings, had provided additional signs that temperatures were rising.[129]

These studies came with an associated body of warnings that focused on the effect this would have on human energy and, most importantly, the American body. Once the mercury rose, Mills predicted, the white American, drained of energy, would be reduced to the savage state found in the tropics. Mills was not alone in this assessment. One anthropologist had gone so far as to suggest that, due to rising temperatures, Americans were turning from blonds and brunettes "to a single type with black and straight hair, high cheek bones and brown eyes . . . in short, we are turning into a variety of humans which is typified by the Red Indian, who was here before us."[130] Others focused on how this temperature shift would dramatically change American cities. E. Woodman, a geologist at New York University, warned his fellow New Yorkers that as temperatures continued to soar the climate would soon be like California, and as he pointed out, being in the "horse latitude," no great civilization has ever grown in California.[131] One Brooklyn journalist even worried that New York might "slip into lazy zone or something worse."[132] "Fossilized umbrellas and fossilized flivver limousines, dug up on the site of New York in ages to come," would only stand as a marker of the terrible weather that impacted the region in the twentieth century.[133] Others projected more optimistic scenarios about what the rising temperatures meant for society. One article predicted that rather than leading to a gradual decline in civilization, the rise in temperature would initiate the dawn of a new utopia.[134] Naysayers busy predicting the decline of civilization had conjured up a catastrophic scenario in which with the rising heat "mankind will be like the Samoans—peaceful, happy, but profoundly lazy."[135] "Machines will run down, stop, rust, and buildings will crumble, never to be replaced, until the whole world becomes a steaming jungle peopled by naked, shiftless savages. Civilization will perish with storminess."[136] In contrast, Hoke put forward a more positive scenario, describing how the movement toward a tropical climate would usher in an era of plenty.[137] In Hoke's utopia, once plenty was the norm, greed, like the ice fields, would lessen.[138] Life would be one long party: "The afternoon draws on. Friends drop in across an ocean for an hour's chat, lovers span a continent to their rendezvous, it is the time for being together, for companionship."[139]

Even while optimistic observers like Hoke predicted the birth of a tropical utopia, fears about a warming climate were tied to wider concerns about racial degeneracy and the draining away of human energy. This speaks to how early concerns about a warming climate were assimilated into cultural narratives about the preservation of American civilization, its geopolitical position, and its racial superiority, as many feared the nation's energy to be waning. Doomsday scenarios described how—as the climate got warmer—the nation's energy would decrease, leading American civilization to resemble the "inferior" races that lay both beyond and within its borders. Mills would cement this view in the public mind, describing how rising temperatures diminished the energy of Western nations, bringing them in line with Asian countries such as Japan and China. In an article published in *Science* titled "Oncoming Reversal of the Human Growth Tide," Mills repeated these arguments, explaining how it had become apparent that "earth temperatures have been trending upward again now for eight decades, but perhaps reaching the critical level necessary to produce growth depression only with the markedly abnormal warmth that has prevailed since 1929."[140] This was reported to the public in the popular press, where Mills described the "wobbly, downward trend" identified in the heights and weights of sixty-five thousand freshman college students. The age of first menstruation, he explained, had begun to arrive progressively later, in line with hotter temperatures and signaling a draining away of both climatic and human energy. If the world continued to get warmer, he warned, "a shorter, sicker, slower population" will result.[141] "Thermometer and yardstick," the commentator mourned, "seem indeed to demonstrate that civilization has again reached one of its several noons and is gradually verging toward another lovely sunset."[142]

Conclusion

By the 1930s, climatic energy had become part of a wider social discourse on race, immigration, and the continued survival of the American nation. American exceptionalism, as determinists like Huntington and Mills never failed to point out, was the result of its energy, and many believed that its stability depended on maintaining, controlling, and managing the energy on hand. Not only did this manifest itself in potent threats from outside (the misdirected energy of immigrants being a prime concern), but as the climate appeared to be warming, the United States appeared only ever a thermal degree from collapse. And yet, while many looked to climatic energy as an external measure for society, others stressed the importance of direction. After all, as Tchijevsky recognized, because energy "only obliges us to do *something*," what that activity would be was difficult to predict. Energy could lead to war, or the flourishing of creativity and artistic endeavors. As the problem of "misdirected energy" showed, more important than the amount of climatic energy on hand was the design of the moral and physical institutions through which this energy flowed. Energy thus pivoted awkwardly in this paradox, standing as an external measure for society, while also demanding a strong social edifice to harness and assimilate this energy to socially productive ends. Furthermore, as we shall see in the following chapter, it remained unclear what came first: was American society an actualization of a latent energy in the environment, or was it dependent on its capacity to channel energy into form?

CHAPTER THREE

Energizing Culture

Psychic Energy and Primitive Waterways

> All matter and energy is a gift. No one has created it, no one has earned it, no one deserves it, and no one therefore should be allowed to appropriate it selfishly. Man's economic function is simply to wrap this gift in convenient parcels. His own ability to do this cleverly and effectively and the tradition of doing this cleverly and effectively, is also a gift.
>
> —*Lewis Mumford, "Personal Note," 1918* [1]

A year after arriving in the United States in 1940, the Austrian psychoanalyst Wilhelm Reich was camping in New England when he looked up at the sky to see it shimmering.[2] Reich was convinced that the twinkling detected in the night sky was orgone, a physical manifestation of the libido—a form of psychological energy famously proposed by Sigmund Freud as early as 1894. Orgone, Reich ascertained, was a life force, a mysterious current able to both break down sexual repressions and cure a variety of physical maladies, including cancer.[3] Reich's identification of orgone outside the physiological organism concluded two decades of research into whether the libido could be materialized. Not only was this of scientific importance, but it also added support to Reich's long-standing attempt to marry psychoanalysis with Marxism.[4] Orgone suggested the existence of an a priori energy independent of the socioeconomic structure. Only once culture was in harmony with its energy could human freedom be realized. True democracy depended, Reich maintained, on the awareness that "the inner and not the external law is the yardstick of genuine freedom."[5] Now looking up at the night sky, Reich was convinced that orgone had an objective reality in the universe.

Even though many suspected orgone to be a product of Reich's imagination—a sign of worsening mental health, paranoia, and even possible schizophrenia—he had not been alone in seeking a new energy basis for society. In fact, the

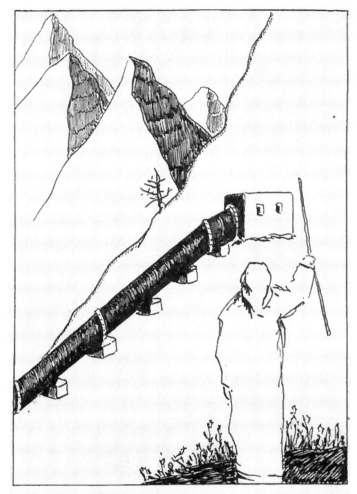

If Moses should smite the black rock?

Illustration of Giant Power. R. Bruère, "Pandora's Box," *Survey Graphic* 51, no. 11 (March 1, 1924): 559.

appeal to rehabilitate society's energy was not confined to psychoanalysis but surfaced in other disciplines concerned with the built environment and physical power. Four years prior to Reich's siting of orgone, Lewis Mumford, one of the nation's leading intellectuals, called for the international power industry to repair the synthesis between energy and culture.[6] Mumford made a case for this at the Third World Power Conference held in 1936 in Washington, DC, a high-profile event bringing together the world's leading experts in fuel and power.

In this speech Mumford discussed the relationship between energy and culture, drawing on the three phases of social evolution outlined in his book *Technics and Civilization* (1934). The eotechnic phase in Mumford's schema stood for a "happy, free, Arcadian existence" when man was in harmony with his energy. In contrast, the paleotechnic period, which represented the rise of industrial society, was symptomatic of a pathological society, as energy was sublimated into unhealthy channels by a culture dominated by pecuniary values.[7] During this period, Mumford told his audience, power had been appropriated by predatory groups "that seek in the manner of Procrustes to make the activities of the communities fit their narrow and arbitrary pattern."[8] Because of this maladaptation, the paleotechnic period was the result of an unhealthy sublimation of energy through a repressive social structure. To counter this paleotechnic malady, Mumford called for a revolution in technics—a return to an energy base aided by technology that could rival the sexual revolution later advocated by followers of Reich.[9] Mumford thus gave a stern warning to the power industry: at the present moment only a modicum of the free energy from the sun was transmuted into the "material and immaterial instruments of human culture—handsome cities, comely landscapes, works of literature, art, science, and above all, communities of well-endowed men and women who are in full possession of their social heritage."[10] To cure the current disjunction, Mumford proposed two alternatives: the power industry must provide "either a comprehensive collective effort to transmute power into culture, or a giving way before those forces that would pervert human values, debase life, and finally, in a paroxysm of brutal rage, destroy themselves in the act of destroying civilization."[11] Despite his social marginalization, then, Reich was not alone in advocating the return of culture to an energy base. Instead, there was an increasing awareness that the United States, with its poorly planned cites, energy infrastructure, and transport networks, had fallen victim to the same false relation engendered within a pathological psyche.

This chapter will uncover how a discourse on *psychic energy* surfaced during the interwar decades as part of a wider debate about the revitalization of American culture. It will trace how an energy language drawn from modern psychology filtered beyond the psychological establishment and a niche countercultural movement to influence the work of cultural theorists and planners, and even informed how one of the nation's leading infrastructure projects, the Tennessee Valley Authority (TVA), was envisaged. Scholars such as David Nye, Bob Johnson, Ronald Tobey, and Jennifer Lieberman have demonstrated how energy, and in particular electricity, became central to a progressive reimagining of American society during these decades.[12] The fact that much of this rhetoric was also grounded in an energy language informed by modern psychology has until now been largely overlooked. As this chapter will show, however, physical power was often described via energy metaphors drawn from the mind. Psychic energy, we shall see, provided a powerful model for understanding the built environment and the progressive power projects that came to define the period.

The chapter will start by tracing how psychoanalysis was received in the United States in the 1920s and how this led to the circulation of a new discourse on psychic energy within American thought. It will tease out how new psychological models of the mind provided two competing ways of understanding the relationship between energy and society. One, best represented by Reich's Marxist agenda, sought to align American society with an a priori energy latent in the environment. The other, emerging out of Mumford's politics of rejuvenation, looked to emulate Freud's therapeutic method and shape and redirect the flow of energy within American society. Both approaches, we shall see, became central to the anti-modern critiques that ran through the decade, as figures ranging from Oswald Spengler to D. H. Lawrence and the Regional Planning Association of America (RPAA) were drawn to "primitive" energy as a way of reinvigorating American society. As psychic energy was conflated with physical energy, moreover, it provided a shared framework to imagine the built environment and progressive energy projects such as Giant Power and the TVA. The polymath planner, member of the RPAA, and conservationist Benton MacKaye, for example, believed that American culture had to reconnect with an "indigenous" energy, or a "psychic resource" located beneath the web of industrial civilization. MacKaye understood the TVA in these terms, describing it as the mapping of American culture onto a resource base, following the contours of the watercourse rather than the arbitrary boundaries of states' rights.

Once again, the desire to bestow ethical value on energy was subject to the same internal contradictions that had run through the debates on progressive energy and climatic energy. As planners and intellectuals sought to locate an energy in the environment as a regenerating force, they confronted the energy paradox: could one locate a "primitive" energy in the environment as a blueprint for society, or was energy only granted moral value as it was harnessed into social form?

Reinterpreting Freud: Lewis Mumford's "Golden Day"

In 1926 Geddes Smith, managing editor of the *Survey*, wrote a letter of congratulation to Lewis Mumford.[13] He wanted to praise him on his recent work *The*

Golden Day (1926), which, he had to admit, even a journey across North Carolina in an overheated Pullman could not ruin. Smith, along with Mumford, was involved in the RPAA, a group that had coalesced around the architect and planner Clarence Stein following a meeting in 1923 at the Hudson Guild Farm in Mount Olive, New Jersey. The group, which included Charles Whitaker (editor of the *Journal of the American Institute of Architects*), the forester Benton MacKaye, the economist Stuart Chase, the businessman Alexander Bing, and the architect Henry Wright, among others, had set out to explore the nascent field of regional development.[14] Interested in the Garden City movement promoted in England by Ebenezer Howard, the RPAA sought alternative planning models to the congested, blighted "dinosaur cities" of America.[15] The RPAA was far from alone in calling for the redistribution of power away from urban centers. Indeed, growing interest in decentralization was reflected by the rise of regional movements across the country, from the "Southern Agrarians" (a collection of writers based at Vanderbilt University) to the craze for "going Borsodi" (the return to traditional homesteading).[16]

In his letter, however, Smith was busy thinking about Mumford's book and how it related to his recent study in psychoanalysis. He had been trying to get at the social significance of the synthesis Mumford had posited and challenged him to consider whether "that central flow of the libido—you call it by the good old term 'desire'—which will if God is good make me fitter for a part in such a world as you hint at. I know that sense comes in many ways: I have no wish to force your thought into my particular pathway, but I cannot imagine any sufficiently regenerative influence rising in the environment to bring about what you look for. Can you?"[17] Smith was right to interpret Mumford's cultural synthesis in terms of Freud's libido.[18] Like many of his contemporaries, Mumford had moved through a self-directed program of study in psychoanalysis, reading the work of Sigmund Freud, Carl Gustav Jung, Pierre Janet, and Ernest Jones.[19] In *The Golden Day*, Mumford located America's revival in a "palpable past," a golden day found in colonial New England. This golden day was a perfect temporal moment, an age of social and psychic equilibrium. In contrast, Mumford understood modern civilization to be suffering from a pathological malady. Megatechnics (society's superego) had misdirected energy into aggressive and harmful ends associated with the acquisitive instinct. This synthesis had broken down when, Mumford explained, "the energies that should have gone into the imaginative life are balked at the source by the pervasive instrumentalism of the environment."[20] While there was no doubt that the golden day was a time of unhampered libido, the answer to Smith's question remained ambiguous.

Could the libido emerge a priori from the environment, or did it have to be molded by society? Was the libido inherently good (and by proxy a valuable model for society), or was it amoral and therefore in need of form?

Smith was not alone in finding Mumford's stance on this obscure, but the psychoanalytic institution was equally divided on the subject. As the field of psychoanalysis was reinterpreted in the United States, the nascent cultural pessimism following the First World War led Greenwich Village intellectuals to celebrate the libido as an escape from the strictures of contemporary society.[21] The spirit of rebellion endemic to this period read the libido as a positive force, a counterpoint to the moralistic nineteenth-century culture that placed punitive restraints on the individual and society.[22] Much to Freud's distress, on the assimilation of psychoanalysis into American popular culture, emphasis had shifted from the rehabilitation of the libido (as performed within Freud's clinical model) to its revolutionary potential.[23] This stress on the liberating elements of the libido echoed Reich's subsequent challenge to Freud's therapeutic method. In Reich's metapsychology, the biological energy embodied in his concept of orgone was granted its own moral priority. He privileged the natural flow of the libido over the contingent structures of civilization, meaning that the remedial onus was on civilization to adjust to this underlying energy. Tying Marxism to psychoanalysis, energy, Reich maintained, was both secondary to social development and intervening in its process.[24] Under the current capitalist system, energy and culture, sex and morality, democracy and politics had become alienated.[25] The rise of Nazism, Reich believed, had revealed the unnaturalness and polarization of modern civilization: its mass spectacle diagnosed as a desire for freedom perverted in the course of achieving it.[26] In Reich's view, civilization had to adapt to, rather than hinder, the libido's natural drive.

In contrast to Reich, Freud had challenged the metaphysical priority of the libido in "Beyond the Pleasure Principle" (1920), where he undermined any teleological directive attributed to energy. Here he posited the cohabitation of two energy drives, one productive (Eros) and the other destructive (Thanatos).[27] In *Civilization and Its Discontents* (1930) he went one step further, arguing that because of the fundamental economy of energy, civilization depended for its survival on sublimating the libido into higher cultural forms. Civilization siphoned off the limited supply of energy from sexuality into productive ends and thereby regulated energy (an undirected force) away from sex into the cultural forms that built a civilization.[28] In Freud's model of the psyche, therefore, the clash between the libido and civilization was irreconcilable; the libido threatened the stable edifice of civilization, just as civilization depended on repressing the power of the libido. Moreover, rather than civilization adapting to a primordial energy flow, Freud placed the responsibility on civilization to rehabilitate this energy to higher ends. This would eventually lead Reich to accuse Freud of writing *Civilization and Its Discontents* as a direct attack on the ethical privilege he had bestowed on the libido.[29]

Needless to say, whether or not the libido was intrinsically good (and threatened by civilization) or bad (requiring recuperation through it), Freud's psychoanalytic model placed energy and civilization in binary opposition. As a result, despite Freud's clinical model undergoing constant reinterpretation (and more often than not misinterpretation), it provided a dualism in which the unconscious contained a reservoir of energy repressed by consciousness, or by civilization. Freud had transformed this into a structural model in "The Ego and the Id" (1923), providing a system-based model of energy, constituted by the id (the innate, primordial drive), the ego (the reality principle), and the superego (the moral censor). As an essential biological reality, the id stood chronologically prior to the evolutionary development of the race and the individual.[30] In contrast, the superego was the moral censor, representative of society's ethical dictates and laws. While the essentialism of the id stood as a fundamental biologic drive, the superego was based in time. The libido, which Smith believed Mumford had unearthed in the golden day, was located in this a priori time, revealing itself in a productive drive for renewal. Echoing the transcendentalist desire to excavate an indigenous American culture, Mumford's "libido" emerged from an eternal moment, an unconscious that preceded the web of civilization.[31] By locating the golden day in a primordial id, Mumford maintained faith in the productive potential of energy. For Mumford, as Gregory Morgan Swer has argued, the libido was "to contain such higher sentiments as love, and the desire for creation and self-expression."[32] As opposed to taking the antagonism between the libido and civilization to be irreconcilable (as Freud did), Mumford took the golden day to be a moment when they were in sync.[33]

Primitive Energy and the Critique of Modernity

Mumford was not alone in identifying a primordial energy in the environment as the basis for social renewal. In fact, the desire to reconnect with a "primitive energy" in the environment became central to a strand of anti-modernist critique that swept through European and American society in the interwar period. This critique was fueled by the writing of the German historian and philosopher

Oswald Spengler. To Spengler civilization could not be aligned with an essential life being, and it was this fissure that would eventually force civilization to collapse. Spengler outlined this thesis in *The Decline of the West* (1918), a bestseller in Germany and America.[34] In this work, Spengler envisaged civilization as a living organism, moving through the stages of spring, summer, autumn, and winter. This life cycle was driven by the opposition between organic life and artificial civilization. As civilization progressed, the gap between primordial man (living in his environment) and "thinking-being" (embedded in cities) widened.[35] "Primitive man," Spengler maintained, could be understood as a form of "plant-life," preserving a connection to cosmic being and the ebbs and flows of the solar cycle. Spengler even took the unique connection between sunspots and war to be an example of how primitive life stood in communion with a throbbing universe.[36] The more man progressed along the route of civilization, developing mechanisms of independent thought, enforcing the division of labor, and building up metropolitan centers, the further abstracted man became from "plant-life" and the cosmic unity at work. Civilization's decline occurred when this cleavage intensified to such an extreme that artificiality at its limits could no longer sustain its growth. Now in the final stage of this process, modern society, Spengler recognized, had become entrenched in urban centers, economic institutions, and legal and political systems, which had subsequently extended arbitrary structures over organic life. Liberal democracy, as it was celebrated within the United States, epitomized the end point of this schism: the legitimization of multiple viewpoints, undermining the premise that one essential truth or natural order existed. Outlining many of the arguments that later found a home in fascism, Spengler popularized a view of "Faustian culture" so entrenched in an instrumental, artificial, and mechanized order that it was on the brink of collapse.

Mumford drew on Spengler's paradigm of "Faustian" culture, sharing his view that civilization had reached the final stage of artificiality before its collapse. However, rather than upholding the binary between civilization and organic life, Mumford was drawn toward the potential for synthesis.[37] This possibility was not included in Spengler's text, but it was how members of the RPAA translated Spengler's cultural pessimism. For example, Benton MacKaye, one of the most active members of the group, absorbed Spengler's diagnosis that Faustian culture had entered winter, concentrating into a "city, a point."[38] Yet while the city was the last stage for Spengler, MacKaye looked forward to a period of synthesis, when the polis would actualize a "natural resource" located in the indigenous environment and transform it back into "culture."[39] By 1939 the implications of Spengler's pessimism had become apparent, as Mumford attacked him for popularizing his cynical dualism and sanctioning a descent into "barbarism."[40] Mumford felt that Spengler's binary had legitimized the rise of the Nazi Party in Germany and fascism in Italy and had given ideological credence to "their irrationalities, their phobias, their humorless limitations, their colossal brutalities, their perverse animus against all life, except at the blindest levels of the id."[41] Spengler's shortfall, Mumford felt, was his disregard for the human itself, his hatred for "the independent power of the mind, creating values, erecting standards, subduing ferine passions, laying the basis for a more universal society."[42] By making the cleavage between nature and civilization inevitable, Spengler, Mumford felt, had eliminated hope from the equation and, as a consequence, any faith that this cleavage could be repaired.

In contrast to Mumford, those attracted to Spengler's fatalism had sought to reconnect to this organic sphere, seeking out a primordial energy latent in the environment.[43] Energy, from this perspective, was regarded as a counterpoint to artificial structures that had wrenched the human from his instinctual drives. Energy provided an essential framework—a primordial id underneath the industrial web. The fashion for outdoor leisure, sun worship, folk culture, and organic farming that emerged in Britain and Germany after the First World War drew on this rhetoric, locating a primitive energy at the heart of an anti-capitalist and anti-modern critique.[44] Elements of this had already been found in the German Monist League, as Ernst Haeckel proposed sun worship as the foundation for a modern religion. Because the sun was the origin of all life, sun worship, Haeckel believed, was "a form of naturalistic monotheism . . . a much better foundation than the anthropistic [sic] worship of Christians and of other monotheists who conceive of their god in human form."[45] The same symbolism spread through the outdoor movements, with many, inspired by Spengler, turning to energy as a primordial entity, an essential being opposed to the contingency, relativism, and artificiality of modern life.[46] These trends enacted a form of social recapitulation, reengaging with a trans-temporal, cross-cultural, primordial humanity.[47] From the adulation of the sun in heliotherapy to an electrifying communion activated through Morris dancing, energy provided access to this other temporality.[48] The British youth movement the Kindred of the Kibbo Kift, for example, dressed in an assortment of Anglo-Saxon, Celtic, and Native American attire, began their annual gathering, known as the Althing, by pledging allegiance to energy. The fire keeper would chant,

> Energy, energy, ceaseless energy!
> The silent and terrific energy of the Universe!

The faithful and wonderful energy of the Electron!
Macrocosm and Microcosm!
One,—One,—One,—One is One!
All is Energy,—the Energy of One![49]

For members of the Kibbo Kift this ceremony provided a symbolic gateway from an increasingly mechanized, deskilled, anonymous modernity into an eternal realm infused with folk legend. Through ritual, energy became a potent cultural symbol, offering a temporal escape from civilization back into an ancient past.

The British writer D. H. Lawrence did the most to cement this imagery. Lawrence tied energy to an anti-modernist critique, conjuring up a powerful "blood-consciousness" that flowed through society, a primordial energy that bypassed rational thought to act directly on the body. In his short story "The Sun," for instance, a young woman embarks on a love affair with the sun, a relationship that takes her outside time, back to the fall of man. Here Lawrence captured how the sun pressed down on her: "His blue pulsing roundness, whose outer edges streamed brilliance. Pulsing with marvellous blue, and olive, and streaming white fire from his edges, the sun! He faced down to her with his look of blue fire, and enveloped her breasts and her face, her throat, her tired belly, her knees, her thighs and her feet."[50] Because energy was associated with the eternal and the uncorrupted, Lawrence chose America, rather than Europe, as the base for his social and sexual revolution. Due to America's "relative youth," it had a deeper connection, Lawrence believed, to its primordial energies; it was a land saturated by "blood-consciousness."[51] In his *Studies in Classic American Literature* (1923), which read much like Mumford's *The Golden Day*, America was held as a vessel of primitive energy, home to the "Red Indian," to "blood-consciousness," to "vibration." Lawrence described how in America the "human apparatus," like the "wireless," goes "tick-tick-tick" and receives "all kinds of messages.... Only the soul, or the under-consciousness deals with these messages in the dark, in the under-conscious."[52]

Just as Lawrence hoped that reestablishing a connection to primitive energy would rehabilitate a weakened race, he also saw it as a means of reactivating a limp democratic system. He was not alone in this sentiment. Indeed, acolytes like Rolf Gardiner, who had led the rural revival in Britain, called for a centrifugal force to activate a wilted democracy. Gardiner looked for inspiration in Germany's youth movements (the *Wandervogel*), in folk dance, and in organic farming as a means to enact a greater social communion.[53] These activities fed into his political tract, *World without End: British Politics and the Younger Generation* (1932), published the year Hitler assumed power in Germany. Drawing on Spengler, Gardiner characterized the party political system, along with liberalism, as the final point before civilization's collapse. Countering this with a metaphysics of force, Gardiner celebrated the dictator as "an epicenter from which energy would radiate."[54] Like the revolutionary leader in Lawrence's *The Plumed Serpent* (1926), who started an uprising by radiating energy from a nodal point, Gardiner was certain that what was needed was a centrifugal energy to regenerate society: "Someone has got to fling a pebble into the stagnant waters of our imaginative life and cause the circles to ripple."[55] Gardiner therefore called for "a form of *comitatus*" similar to an elite "*Bund*," based on "common experiences, common forms, common hopes," "a ferment which fructifies dead material, which impregnates all that it comes in contact with, with fertilizing energy."[56]

While the associations that clung to this primordial energy found their logical expression in a defense of fascism, with a central dictator energizing the masses, liberals were also inclined to see in this primitive energy a means of invigorating social democracy. Benton MacKaye recognized this, noting strong affinities between the folk revival in Europe and the core objectives of the RPAA. The similarities between the two groups, he argued, were found not in the particular methods they deployed but in the novel way they were dealing with modern problems.[57] He mentioned this to Mumford in 1924, as he recommended a book on the Kindred of the Kibbo Kift, which Clarence Stein had brought back from Europe.[58] What interested MacKaye about the Kindred of the Kibbo Kift was that their project resembled his own model of the Appalachian Trail. Rather than replicating the past, they were using it as a means to find a fuller way of living. Thus, while the RPAA incorporated similar folk rituals at their annual gatherings, Mumford warned against a simple emulation of folk culture—a problem he had begun to detect in the European folk movement and that had been brought to light by Henry Ford's kitsch preservation of the Wayside Inn.[59] Rather than dredging up the past through masquerade, what was needed, Mumford believed, "is not a revival of the old: it is a fresh growth of something new, similar in animus and method, at times, to what has existed in the past, but with all the differences that the intervening time has wrought . . . simplicity and the primeval life have their permanent place in the soul, even though we use turbines instead of water-wheels."[60]

Waterways of the Mind

The appeal to a primitive energy, therefore, provided a schematic view of civilization as a layered entity like the psyche. This worked both ways, with the psyche understood topographically through the features of a landscape. This shared analogy meant that psychic energy became just as easily a way to describe the movement of energy through the physical environment as through the mind.

Sigmund Freud had long utilized geographical metaphors to describe the psyche, referring to the libido as a river or stream that got blocked and overflowed.[61] William James had used similar language, filling his writing with images of "rivers," "dams," "streams," and "free water."[62] Reich himself echoed this terminology, as he described the libido as a river under threat from "artificial dams" that impede its natural current.[63] Although geographical metaphors had long saturated descriptions of the psyche, the connection between the mind and the natural environment became most explicit in the work of the British botanist Arthur Tansley. Tansley was a lecturer in botany at the University of Cambridge, had become well-known in the scientific community as the founder of the journal the *New Phytologist*, and was appointed president of the British Ecological Society in 1913. Yet despite his emerging career as one of the leaders in the field of ecology, Tansley turned to psychology in 1920 with his book *The New Psychology and Its Relation to Life*. After a favorable review by Ernest Jones, this book would be a great success, selling over ten thousand copies in the United Kingdom within three years of its publication.[64]

Tansley interpreted Freud's metaphor of the mind literally as a hydraulic system, describing the psychic apparatus according to the features of a landscape, conjuring up the "channels" through which psychic energy flowed as if it were a river system.[65] These channels, Tansley described, hovered between "*universal complexes*" and "*particular complexes*."[66] Complexes, Tansley argued, could be understood as a "channel," "in its primitive meaning of the water course."[67] He elaborated: "Water, originally possessing the potential energy of position, flows along a channel and does work: it either cuts its channel deeper, or it may be made, for instance, to turn a mill wheel. If the flow is dammed by an obstacle the water banks up behind the dam and acquires a new 'head.'"[68] Explaining how "the behaviour of the energy of flowing water gives a fairly close picture of the behaviour of energy in a conative channel," Tansley extended Freud's hydraulic metaphor to materialize the libido in the contours of the watercourse.[69] Here Tansley's natural metaphors of the "watercourse"—animated by "flowing water," "dams," "channels," and "stagnant" areas—evoked a topography in which there existed a layering of cognitive channels with energy flowing "along the line of least resistance."[70] The route of least resistance was that of "primitive conation,"

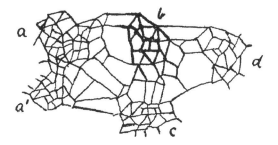

Illustration of the associative network of the mind seen from above. Five complexes are being illustrated. B, which is in bold, is shown to be in a state of activity. A. Tansley, *The New Psychology, and Its Relation to Life* (London: George Allen & Unwin, 1920), 48.

whose plan was laid down by heredity and developed by a "normal environment."[71] Normal psychological development depended on conjuring up "psychic energy" to carve out new connotative channels in the mind.[72]

When Mumford read *The New Psychology and Its Relation to Life* in 1921, a year after it was published, he found Tansley's description of the mind as a watercourse appealing, not least because it could be understood through his mentor Patrick Geddes's model of the "Valley Section."[73] Seen from a bird's-eye view, the difference between a metaphorical and an actual watercourse was eliminated, and the flow of psychic energy could be substituted for physical power. Mumford had encountered Geddes's work in 1914, and he adopted his structure of the valley section, with its motto of "place, work and folk" as a guide for his regional planning philosophy.[74] The valley section spoke to the vital connection between man and his environment, to his engagement and interaction with the valley milieu.[75] By the time Mumford read Tansley's book, he was well acquainted with the recent trend in collating surveys, maps, and charts that had emerged from Geddes's geographic school. A year earlier he had spent five months at the home of the regional survey movement in London, Le Play House, editing the *Sociological Review*, despite Geddes being away in India at the time.[76]

What caught Mumford's attention in Tansley's book, nonetheless, was a section titled "By-Ways of the Libido."[77] Here Tansley discussed how psychic energy was diverted through conative channels following the line of least resistance.[78] These channels were in a state of flux, reliant on the stability of the psyche. What attracted Mumford, as he copied this section from Tansley's book into his personal notes, was how when this adjustment occurred through the subversion of the end point of conation, a modification in the primary interest of the affect followed.

Taking the example of the truncation of the sex instinct, Tansley explained how "the libido, unable to pursue the conation to its proper end, seeks hither and thither for an adequate outlet."[79] The capacity to redirect the end point of conation (shifting affection from certain drives) provided a diagrammatic model of how energy could be channeled through the psyche in the same way as it could be diverted through the physical environment of the watercourse.

Mumford later recognized this as a powerful vision for the planner as much as for the psychologist. In an article published in the special issue of the *Survey Graphic* dedicated to the Regional Planning Movement in 1925, Mumford redefined planning along these lines.[80] Tracing three different stages of mass migration within the United States (leading to the most recent influx of population into concentrated urban centers), Mumford determined that America was in the "midst of another such tidal movement of population."[81] He termed this the fourth migration: the movement of flow back into the hinterland, enabled by new ways of harnessing power, innovative channels aided through the capacity to generate electricity, and the rise of motor transport. With a novel ability to shape the direction of this flow, Mumford argued, "we may either permit it to crystallize in a formation quite as bad as those of our earlier migrations, or we may turn it to better account by leading it into new channels."[82] Able to wrench open new channels and redirect flow, Mumford described the planner's role along the same lines as the analyst: he was "to suggest what these new channels are, to show how necessary it is for us to trench them open, and to indicate how much the future may hold for us if we are ready to seize our destiny and shape it freshly."[83]

Psychic Energy Meets Giant Power

Giant Power, of early interest to the RPAA, was understood by the group in these terms, revealing the slippage between psychological understandings of energy and physical power. It was to sublimate physical power (through the extension of long-range electricity transmission networks) away from the unhealthy complex of the urban megastructure and redistribute it to the regions. Giant Power was a progressive plan put forward by Gifford Pinchot, who, as former chief forester in Theodore Roosevelt's administration, had had a long adversarial history with the private utilities. On his election as governor of Pennsylvania in 1923, Pinchot supported a Giant Power survey to investigate the monopolization of the state's power resources. To do this, he placed Morris L. Cooke, one of the foremost advocates of scientific management, as head of the board. Giant Power emerged as a proposal for the state to plan and control the electricity networks, taking away sovereignty from the private utilities.[84] By the 1920s advances had been made in the transmission of high-voltage current across long distances, enabling large areas previously inaccessible to be linked to a network of supergenerating plants.[85] Despite this new technology, private corporations had refused to extend services, concentrating power in urban centers, where costs were lower. In opposition, Giant Power sought to utilize the new transmission capacity and spread electricity back through the regions. With Pennsylvania dependent on bituminous coal, however, Giant Power replaced river channels with electricity cables to distribute energy derived from coal. Situating large generating plants at mines, high-voltage transmission lines radiated energy through the state and beyond.[86]

The potential to domesticate energy through these "by-ways" provided a basis for regenerating the social, technical, and economic structure of the region. The Soviet Union was a precedent for this, where electrification had been promoted as the foundation of a collectivized economy, embodied in Lenin's famous adage "communism is Soviet power plus the electrification of the whole country."[87] Lenin's desire to melt church bells to provide a light bulb in every village was a powerful vision of culture turned upside down, built on a renewed energy base rather than controlled by a powerful religious ideology.[88] Established in 1920, the Soviet electrification plan, implemented by the State Commission for the Electrification of Russia (GOELRO), followed the principle that a revolution in the energy base would initiate a social and economic revival through centralizing the economy and uniting urban and rural regions. This model had inspired the RPAA, and members within the group had gone so far as to offer their planning services to the Soviet Union in its early years.[89] Although this call went unanswered, within the RPAA Stuart Chase remained the strongest advocate of the Russian experiment. With a "billion wild horses" released by modern technology causing havoc in America, Chase remained in awe of the Soviet Union's ability to domesticate this wild beast.[90] Not subject to misbehavior on the part of the financial institutions, Russia's New Economic Policy had proved that energy could be brought into balance with the needs of "wayfaring men."[91] Taking a bird's-eye view of the flows of energy across the nation, Chase understood Giant Power as a means of eliminating needless waste, of taming this "display of energy and rush" so as to create communities and regions "specifically planned for the maximum of local subsistence and the minimum of cross-hauling—interregional as well as local—communities."[92]

Although some members of the RPAA, such as Stuart Chase, understood Giant Power through the eyes of the Soviet planner trained in efficiency, others focused on the psychological benefits of the plan. In an introductory essay on Giant Power in a 1924 edition of the *Survey Graphic*, Robert Bruère stressed the importance of human engineers working alongside mechanical engineers in the plan for power. The mechanical engineer, he argued, had established systems of distribution only with the acquisitive impulse in mind. On the other hand, social reformers, Bruère explained, had up to this point failed to establish "large plans for the distribution of population and the spiritual enrichment of community life comparable to those which the mechanical engineers have developed for the production and distribution of mechanical energy."[93] A year later, Bruère pointed to the Ontario Hydro-Electric Commission as an example of how this could be successfully achieved. Here the commission halted proposals to place a large suction pump at the falls, thus decentralizing energy by distributing it through small towns. The amazing results, he explained, were illustrated by the local town Woodstock, "growing with its roots in its own ground—not a one-industry mushroom boom town, but a sort of manor town to the surrounding countryside."[94] Had the commission placed "a huge suction pump at Niagara," Bruère predicted, it would have extracted "the life out of their communities and [left] them gossip-rattling, Main Street shells or mere suburban satellites."[95]

Benton MacKaye's "Psychological Resource"

Whereas Bruère stressed the importance of aligning human and mechanical systems, Benton MacKaye went furthest in understanding Giant Power as an extension of the mind. For MacKaye, the environment harbored a dormant infrastructure, a hidden potential waiting to be actualized by the planner. Giant Power, he believed, could map electrical power onto this latent flow in the environment. MacKaye put this idea forward in a memorandum titled "Regional Planning Studies" drafted for the RPAA in 1925.[96] Proposing a number of studies into effective and ineffective energy systems, he cited Giant Power (along with the Deerfield Waterpower Project) as an example of how to avoid the "aimless" and "accidental" structures established by the private utilities, like the Power Transmission System of the New England Power Company. These proposals demonstrated the "planlessness" and the "accidental" models generated by financial institutions that gave little consideration to the underlying topology.[97] The New England Power Company, MacKaye explained, had imposed an arbitrary structure over a latent energy flow. To counter this, MacKaye urged the RPAA to provide a study on Giant Power, so as to present a preliminary inventory of the coal and waterpower of the United States, as the "energy *basis* for a policy of decentralization of the working population."[98]

The importance of returning planning policy to a concrete energy base had not just emerged with Giant Power but had also been central to MacKaye's proposal for an Appalachian Trail. The seed of this project was developed during MacKaye's early career as a forester, working between the US Forest Service and the Department of Labor in Washington, DC.[99] During this period, MacKaye (along with Chase) had been a member of the "Hell Raisers," subsequently following the well-trodden path from the Technical Alliance to the RPAA.

The Appalachian Trail was a proposal for a series of recreational communities to "live, work and play on a non-profit basis."[100] By creating permanent recreational and work opportunities along the route, the Appalachian Trail provided a means of transforming regional resources into human needs and welfare.[101] MacKaye understood Giant Power as an extension of the Appalachian Trail, and

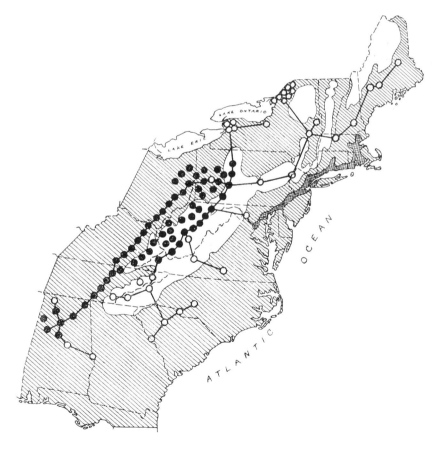

Illustration of Appalachian Power. B. MacKaye, "Appalachian Power: Servant or Master?," *Survey Graphic* 51, no. 11 (March 1, 1924): 619.

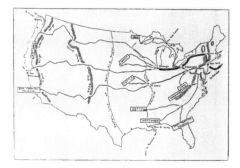
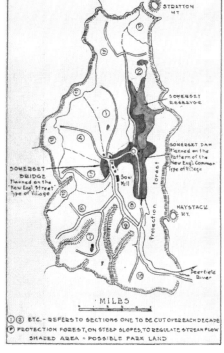

Above, Illustration of industrial flow across the United States. *Right*, The "sphere of origin" in the Somerset Valley, Upper Deerfield River. B. MacKaye, "The New Exploration," *Survey Graphic* 54, no. 4 (May 1925): 154, 157.

an article about their shared objectives was included in the "Giant Power" edition of the *Survey Graphic*. In this article, MacKaye drew on statistics to calculate minimum flow of waterpower and coal in the eastern United States and asked whether this "giant" would be translated into "drudgery and toil" or "creative interest and pursuit."[102] In order to harness energy toward socially beneficial ends, MacKaye revisited the notion of industry, positing two different types of efficiency: "industrial efficiency" ("converting resources into the 'wherewithal' with the minimum of toil") and "social efficiency" ("converting 'wherewithal' into the maximum of 'pursuit,' of 'joy,' of human welfare").[103] Directing this energy toward "joy" rather than "toil" required the infrastructure found within colonial America, rather than the commercial web of industrial civilization.[104] Drawing on Spengler's vision of civilization, MacKaye warned that the commercial web was throttling colonial America. Colonial America could be rediscovered with the use of new technology, but only following a return to that "simplicity" exemplified by "Thomas Jefferson" and "Daniel Boone." The colonial environment,

MacKaye promised, remained latent, and as such, there was a choice: to direct the power into the "commercial tangle" of civilization, or to "upbuild a hinterland in the original modes of purposeful life."[105]

MacKaye understood Giant Power as a means to uncover colonial America, a primordial environment at odds with the web of industrial society. MacKaye's entreaty to upbuild rather than down-build was thus founded on the assumption that there was an a priori channel to build on, that underneath the commercial web was a primitive watercourse, submerged in the environment. Elsewhere, MacKaye described how this indigenous environment would supply the "source of mental life and culture."[106] He expounded on the physical analogy: "as we maintain industry by conserving the natural resources of soil and wood and waterpower, so we maintain culture by developing the 'natural resource' of innate environment."[107] By envisioning the landscape as a layered entity, similar in depth to the psyche, MacKaye redefined the role of the planner: their job was to uncover latent channels, to discover its "hidden potentiality."[108]

In privileging a latent infrastructure under the meshwork of industrial civilization, MacKaye transferred energy into an a priori temporality. Rather than attached to the contingent framework of civilization, this energy represented something eternal and fixed. MacKaye described this layering in his book *The New Exploration: A Philosophy of Regional Planning* (1928). Making allusion to Thomas Henry Huxley's work *Physiography: An Introduction to the Study of Nature* (1877), MacKaye peered down from the Times Building to study the metropolitan flow in Times Square.[109] By doing this, MacKaye compared the water cycle with the traffic that flowed below—the subway trains, shoppers, job hunters, errand boys, loafers, commuters, theatergoers, goods, the ebbs and flows of the city.[110] New York was a "metropolitan wilderness," with tentacles extending across America and across the globe. To counter this metropolitan flow, MacKaye presented a different environment viewed from the top of Mt. Monadnock, a mountain in New Hampshire. From there, MacKaye traced the flows of an indigenous land, an eternal and permanent domain. In contrast to the metropolitan landscape (where "a railway may stop running, or a city may disappear"), in the indigenous land MacKaye saw the "earth itself" as "a receiver and storer of solar energy" that "does not alter: it can never basically alter."[111]

MacKaye described this indigenous environment as harboring material resources, energy resources, and, most importantly, psychological resources. This psychological resource had to be actualized—it had to be translated into culture, just as coal was transformed during combustion. According to MacKaye, the latent energy of the indigenous environment could be released in three different ways: through "mechanical conversion," "biologic conversion," and "psychologic conversion."[112] The first two methods depended on transforming a latent energy, either the sun's rays or energy bottled up as coal, into something evident (into wood, or the moving of a freight train).[113] *"Psychologic conversion"* enacted a similar process, in which a *"base"* was converted into a "reaction." To represent this, MacKaye extended the metaphor of a tree actualizing solar energy during photosynthesis to the "psychologic conversion" that occurred through the creation of an artwork.[114] He explained how "another sort of unseen (or obscure) reality is transformed into substantial evidence. The diffused beauty of a landscape which to the ordinary eye is obscure and perhaps inevident, is, through the brain-action and skill of the artist, captured and placed on sturdy canvas for all to see and comprehend."[115] As the actualization of energy depended on the relationship between "base" and "reaction"—between coal and combustion, environment and painting—culture, MacKaye argued (drawing on Marxist language), should be the reaction on a base, the actualization of a psychic energy. American culture, he maintained, had to be the actualization of a latent energy base, its conversion into cultural form. Planning, therefore, had a clear role: it had "to render actual and evident that which is potential and inevident."[116]

MacKaye was not alone in describing the primordial environment as filled with a psychological resource. In fact, MacKaye had drawn this idea from his friend and colleague, the forester Aldo Leopold. Leopold, like MacKaye, had worked at the Forest Service, where he had been struck by the pervasive backflow of civilization into wilderness regions, brought about by the boom in the tourist industry.[117] Despite having concerns over the invasive nature of this industry, Leopold took the wilderness to be a multifunctional resource, one that contained a high level of recreational value.[118] In an article titled "Wilderness as a Form of Land Use" (1925), published in the industry periodical the *Journal of Land and Public Utility Economics*, Leopold explained why the wilderness was not just the "physical sense of the raw materials it contains" but a vessel for "social values."[119] Once the wilderness had been lost, Leopold argued, so too would "Americanism." "Joliet and LaSalle," he bemoaned nostalgically, "will be words in a book, Champlain will be a blue spot on a map, and canoes will be merely things of wood and canvas, with a connotation of white duck pants and bathing 'beauties.'"[120]

Locating American culture in the wilderness was an idea that extended back to Romanticism and sat at the heart of Frederick Jackson Turner's frontier thesis.[121] This idea had been central to the preservationist arm of the environmental movement led by John Muir, ironically driving the rapid growth of the tourist industry that later threatened this wilderness. Drawing on this tradition, Leopold stressed the premise that the wilderness was a psychological resource, an energy, as multifunctional as coal or timber.[122] What is more, by understanding the American wilderness as a tangible cultural resource, Leopold situated American culture as an energetic reaction. Any threat to the wilderness was a challenge to culture, because without its resource base, American culture, like iron civilization, would be an artificial import, only left with "bathing 'beauties'"—all surface, no depth.[123]

Equally important, however, for both MacKaye and Leopold was the awareness that this psychic resource not only was quantitative but also demanded a qualitative transformation. The tourism industry, for example, failed to actualize this latent energy, as tourists were too busy photographing and commodifying the wilderness, rather than *reacting* to it.[124] When this superficial appreciation occurred, base would remain base and would not be converted into culture. This amounted to a failure to reach the "self-directed evolution" needed for cultural renewal required as a psychological resource was transformed into culture through the act of combustion.[125] MacKaye illuminated how this failed potential could be understood through a comparison of two lovers, Christien and Cyrano in *Cyrano de Bergerac*.[126] In this play one man, Christien, merely appreciates his lover, Roxanne, as a specimen, as "that most beautiful product of natural evolution—a young girl."[127] By contrast, Cyrano actually translates this love, telling Roxanne *how* he loves her, explaining and uncovering this potential. While Christien is all admiration, Cyrano converts—he transforms this latent beauty into art. Translation, therefore, rather than mere appreciation, was central to converting the "latent energy" of the primeval environment into culture. Rather than just admiring the landscape, MacKaye described how one could act on it and transform it into culture.[128] Once again, MacKaye highlighted the fact that it was only through the act of conversion that energy gained social—or, in this case, psychological—value.

It was the threat posed to this psychological resource that led MacKaye and Leopold, along with Robert "Bob" Marshall, to found the Wilderness Society in 1935. The society eventually became the foremost wilderness preservation arm in America, and it was instrumental in passing the 1964 Wilderness Act. In the first edition of their journal, the *Living Wilderness*, the society framed its mission in terms of resource management—psychological energy being the resource under threat. This objective was outlined in the first of the society's platforms:

"the wilderness (the environment of solitude) is a natural mental resource having the same basic relation to man's ultimate thought and culture as coal, timber, and other physical resources have to his material needs."[129] By 1941, this "mental resource" would be suggested as a means of national defense, containing not only cultural but also military potential. With America about to join the Second World War, MacKaye would claim that it should be factored into the "economy of national defense."[130] As this war was a "War of Nerves," the predominant military strategy, MacKaye believed, had to be "the conservation of human energy," not only the mobilization of other resources like coal.[131] Rather than making guns, this resource, he added, "makes *morale*," it enables "neural recreation." It "contained our final reserve of human neural energy—a reserve to be heavily drawn upon—and soon—and increasingly—unless all present portents fail."[132]

The Tennessee Valley Authority and Ol' Man River

The TVA, announced by Franklin D. Roosevelt in his first hundred days in power, would on its inception come to represent for MacKaye the synthesis between energy and culture he had long advocated. A vast redevelopment scheme in the Tennessee Valley, the TVA sought, among other things, to establish a series of dams and transmission centers in a region that covered most of Tennessee and extended into Alabama, Mississippi, Kentucky, Georgia, North Carolina, and Virginia. When understood through MacKaye's regional planning philosophy, however, the TVA can also be seen as an experiment in returning American culture to its energy base—an opportunity to redirect its energy flow.[133] On Roosevelt's announcement of the new authority, MacKaye understood the plan to be as monumental as the revolution enacted by "Copernicus," a physical plan enforcing a "swapping [of] the cultures, not the crudities, of the mountaineer and metropolitan."[134] MacKaye marveled at how the physical plan to control and direct water in the TVA mimicked the cultural one, for to "conserve the basic cultural setting we must control the flow of the metropolis."[135] This, MacKaye believed, could be done in the TVA, through the control of water, electricity, human labor, and population flow. MacKaye even fantasized about how, if he were the "director of public works," he would impose his own model of "townless highways" "to hold in check the 'flood' of population from the cities (even as the river holds in check a flood of waters from the mountains)."[136]

Prior to his appointment, MacKaye was already fantasizing about being the "director of public works." When he was brought into the agency by the southern planner Earle Draper in 1934 (assigned to the Regional Planning section of the

Division of Land Planning and Housing), he confessed that "he had the best job in the TVA and the best job of my life."[137] MacKaye's enthusiasm demonstrates some of the idealism surrounding the organization in its early years.[138] This had been enabled by the vague nature of sections 22 and 23 of the TVA Act, which gave the agency authority "to provide for the general welfare of the citizens of said area"—a clause that expanded federal powers beyond the strict regulation of the river basin to incorporate surrounding areas in an organized regional offensive.[139] Although in the end the clash between the directors of the agency, the idealistic humanitarian Arthur Morgan and the pragmatic David E. Lilienthal, shifted the agency away from the vision of regional planning, MacKaye's employment within the organization demonstrates the visionary potential the agency had on its inception.[140]

Even after MacKaye arrived in Knoxville, he understood the TVA to be a vision inspired by the RPAA.[141] After all, Roosevelt had shown a nascent interest in the regional planning movement and, when governor of New York, had been a panelist at a well-known Regional Round Table at the University of Virginia.[142] However, despite the group's early optimism and the large number of RPAA affiliates (Henry Wright, Robert D. Kohn, Edith Elmer Wood, and Catherine Bauer) employed in the Housing Division of the Public Works Commission, the TVA did not live up to its early promise.[143] Even so, MacKaye understood the role of the agency through his own regional philosophy. Indeed, in a memorandum titled "An Approach to Planning in the Tennessee Valley" he juxtaposed the cultural and the industrial planning aspects of the agency, suggesting how they could be brought together in a synthetic plan.[144] Strained by the bureaucratic nature of the authority, the scope and ambition of MacKaye's vision had little practical influence on the organization, and he complained that the planning branch was stressing the aesthetic elements of regional planning over the importance of "psychic habitability."[145] Preoccupied with beautification rather than habitability, the TVA, MacKaye came to believe, was failing to translate the psychological resources of the region, and his efforts to convince his seniors of this fact remained in vain.[146] MacKaye's most lasting influence was found in an inventory put together by the Department of Regional Planning Studies that archived the "scenic resources" over the 65,000 square miles covered by the Tennessee River, the Valley States, and South Carolina.[147] Throughout this extensive document, the agency did not shy away from utilizing MacKaye's philosophy of regional planning, breaking down the region's "scenic resources" into his designated categories of recreational and wilderness areas, an outline of which was provided in the appendix. Nevertheless, as Arthur Morgan's socially orientated vision came under attack by

David E. Lilienthal, the TVA moved away from its focus on regional planning, and by the time MacKaye left the authority in 1936, his ideas remained largely theoretical projections.

Despite its practical limitations, however, the symbolic potential of the TVA extended far beyond the physical boundaries of the Tennessee Valley. To a depression-ridden nation, it had stood for cultural revival—a reborn America. This image was translated internationally, as many came to understand it as a counterpoint to fascist and communist regimes—an example of social democracy in action.[148] This undermined the criticisms made by those opposed to state intervention, who charged the TVA with being a Bolshevik import. To the essayist Odette Keun, born in France to Dutch parents, the TVA represented "a model" for liberals everywhere.[149] In a book dedicated to the TVA, Keun addressed the "tolerant, socially-motivated, fair-minded men and women" who advocate "a more practical equalization of the chances of health, education, prosperity, and individual liberty."[150] The TVA, Keun believed, had managed this balance, establishing "an equilibrium, within the limits of the capitalist economy, between inimical but not completely irreconcilable forces"—a third way between the evils of capitalism, business rule, and state planning.[151] Echoing this sentiment, Julian Huxley went so far as to suggest that "the initials TVA" had become synonymous with "democratic planning."[152] To Huxley, it was the largest regional planning organization to operate "on the democratic principles of persuasion, consent, and participation."[153] The TVA, he insisted, "has succeeded in demonstrating that there is no antithesis between democracy and planning, and that planning cannot only be reconciled with individual freedom and opportunity, but can be used to enhance and enlarge them."[154] Elsewhere, he adopted the popular "American expression" *yardstick* to propose that the TVA was a yardstick not only "for power production but also for the whole planning of an area."[155]

Not only did the TVA stand as an experiment in refiguring the nation's energy basis, MacKaye later argued, but it also revealed the role of "Ol' Man River" in determining America's political constitution.[156] In a series of articles published in the *Survey Graphic*, MacKaye illustrated how the TVA was a practical example of the fact that "Ol' Man River" was the driver of history. MacKaye compelled his reader to imagine a counternarrative to American settlement, a history driven by the laws of gravity—by energy—rather than by Uncle Sam. What would the country have been like, he asked, had it been colonized "northward-ho" up the banks of "Father Mississippi, and his constituent children—the Ohio, the Missouri, the Tennessee, and his other Tributaries," as opposed to "westward-ho"?[157] "Such common sense cooperation with the law of gravitation," MacKaye insisted, "would have saved a host of headaches for the federal courts, howls in the halls of Congress and barrels of ink in editorial offices. Every state would then have been just naturally a 'Valley Authority' in itself. 'State rights,' so far as affected by stream flow, would be in harmony with the inexorable rights of Ol' Man River, and not be ever at outs with him—as badly as King Canute was purported to be with Ol' Man Neptune."[158] The revolutionary aspect of the TVA, on this count, was that it respected the boundaries of the watercourse—ignoring state borders—rather than following the artificial boundaries imposed by Uncle Sam. Julian Huxley reiterated this, arguing that the flow of water through Mississippi, Alabama, and South Carolina had made a mockery of the arbitrary cartography supporting states' rights.[159] The legal duty of the federal government to maintain flood control and the navigability of rivers for interstate commerce, Huxley explained, meant that a revolutionary form of political power had been forced on the nation by a river. Moreover, the unique form of federal integration affected by the valley had already been extended into international planning agencies that looked beyond national sovereignty, just as the TVA undercut and transcended "State's rights."[160] An example of this was the Danube Valley Authority, a proposal to consolidate peace by integrating multiple European sovereign nations.[161] Yet where the TVA was dealing with a federal state, the Danube Valley Authority had the additional challenge of integrating different sovereign nations, "with different languages, cultures, currencies and economic systems, inheritors of different national traditions and equipped with separate national armies."[162]

While Huxley was more doubtful as to the power of "Ol' Man River" to integrate sovereign nations, MacKaye was confident that this process was inevitable. He believed that "for the obvious good of its inhabitants it is patent that the flow of Ol' Man Danube should be managed as a single fluvial unit."[163] What is more, MacKaye argued that, as the flow of energy extended beyond the watershed to the hydrological cycle circulating through the biosphere, the integration of the entire globe would eventually be necessary. Writing in 1951, as a fully-fledged member of the world federation movement, MacKaye understood that planning had to extend beyond the limited boundaries of region, as he turned his attention toward extending his regional philosophy in the direction of postwar internationalism.

Conclusion

Psychic energy, as this chapter has shown, therefore remained a concept far from limited to the psychological establishment, circulating as part of a wider debate around the revitalization of American society. As a metaphor, psychic energy collapsed the mind with the physical environment and, as a result, provided a blueprint through which to understand the progressive power projects that shaped the period. As cultural theorists and planners struggled to pin down energy's ethical value, however, consensus was once again split about where this should lie. Should America's cultural revival depend on a therapeutic reshaping of energy within industrial society? Or should American culture be aligned with a primitive energy located in the indigenous environment?

Psychic energy, therefore, remained subject to the energy paradox. Rather than supplying a privileged model for American culture, when its logic was pushed to the extreme, it would eliminate cultural difference in its wake. After all, as MacKaye applied his regional philosophy to visions of a world state, rather than providing a psychological resource for an indigenous culture, energy would amalgamate the diverse cultures of the globe. In breaking through the watercourse, coursing over dams, evaporating into the atmosphere, circling the globe, energy would eliminate the indigenous culture its logic presupposed. Far from providing a valuable psychological resource for culture, when culture was synced with an energy base, cultural difference would be eradicated in the flood of "Ol' Man River." Energy, as a universal substance, would be no different from the meshwork of iron civilization as it extended from the metropolitan centers across the globe. Though primitive energy had been essential to conceptualizations of an indigenous American culture, energy in the same move would eliminate cultural difference altogether. As we shall see in the next chapter, when its logic was pushed to the extreme, energy could reduce the entirety of social and cultural life to one common standard.

CHAPTER 4

Erginette's Beauty

Energizing the Economy

Erginette was a beauty. Described by the writer John Erskine in a short story published in *Collier's* in 1933, she was held up as "the example and justification of the Technocratic State, offspring of forethought, flower of regimented habits and scientific diet, as thrillingly at peace with her universe as a spinning dynamo."[1] Like her namesake and idol the erg, Erginette was an ideal energy form. As she grew up "she worshiped Technocracy with every ounce of her vibrant youth . . . yet she pushed her worship to the point where she seemed to be praising the wrong thing."[2] As she became the mirror of perfection, performing energy better than energy itself, Erginette wreaked havoc on the Technocratic State. Overwhelmed with sexual desire for this faultless energy form, suitors gifted "erg" rations to the young lady, diverting energy from "innocent homes" toward the exploration of love.[3] With her value quickly inflating through the bequest of her admirers' energy rations, Erginette took on the semblance of money. Sexual desire would upset the balance of the ideal energy system, as the impeccable energy form became a usurious clog in the system. In Erginette's case, lust after the perfect energy form would transform energy back into money.

Erskine's satire, published just after the Technocracy craze had swept through the public imagination, mocked the energy theory of value that had recently garnered so much attention.[4] During Technocracy's moment in the spotlight, energy, like Erginette, had surpassed its identity as physical form to be fetishized as a cultural ideal. As the Great Depression entered that "historically cruel winter of 1932–33," energy emerged as a new social standard.[5] The head of the New York Stock Exchange, Charles N. Edge, even joked that "'a new pretender to the throne' had arisen. Is it inevitable that energy, the 'god and gold' of technology, be accepted as our control and future overlord?"[6] It was not surprising that energy held such appeal during the winter of 1932. The results of the Energy Survey of

Illustration of Erginette, which came with the accompanying caption: "If women upset the Garden of Eden, what couldn't they do to the engineers' perfect State?" J. Erskine, "Erginette: A Future Helen amid the Topless Towers of Technocracy," *Collier's Weekly*, March 25, 1933, 10.

North America, conducted by the Committee on Technocracy at Columbia University, led by Howard Scott and Walter Rautenstrauch, had verified that there was no shortage of material wealth in America.[7] On the contrary, thanks to technology, energy had never been so abundant. And yet, in spite of the influx of energy, a nation was struggling without clothes, food, and shelter. Factories remained shut, crops rotted in fields, money piled up in banks, and houses were foreclosed.[8] Two years later, Stuart Chase looked back and recognized that 95 percent of the nation's energy had been going to waste.[9] The "energy, the capacity to do work is here, a living demonstrable reality, but it has not got through to the wayfaring man as expressed in his standard of living; the essential work has not been done."[10] The price system was abstracting man from his energy, and the problem was as much psychological as physical. Diagnosing the state of the nation, the journalist Gerald W. Johnson pronounced, "The energy of the country has suffered a strange paralysis. We are in the doldrums, waiting not even hopefully for the wind which never comes."[11]

Even though Technocracy had largely been discredited by the time Franklin D. Roosevelt was inaugurated as president in March 1933, the promise of energy did not disappear in the early years of the New Deal. Instead, it saturated the nation's political and cultural imaginary. Movements from both the political Left and Right began to draw on the language of energy to legitimize their own political programs. Energy became entrenched in radical reform movements, resonated through anti-Semitic rhetoric, was legitimized within New Deal policies, and was adopted by anti–New Deal liberals. Poets and artists alike drew on the promise of energy, seeing in it a way to rebalance a social, creative, and political life ruptured by money. By 1933, energy had almost become akin to a spiritual creed. Writing in *Scribner's Magazine*, Virgil Jordan noticed that, with a price system wrapped in the "language of original sin" and Scott playing the role of a "Technocratic Christ," the new "trinity is the erg, the electron and entropy. Energy is the jealous Jehovah. There is no God but the kilogram calorie and the engineer is his prophet."[12] With "money, the root of all evil," on one side and the "erg" on the other, Jordan concluded that "the business cycle is simply the old story of the battle between the great god Erg and Mammon, the Almighty Dollar, ever in irreconcilable conflict like Ahriman and Ormuzd, Gog and Magog."[13] Opposed to Mammon, energy became a new hard rule, a "yardstick" valued for the characteristics that money lacked: intrinsic value, objectivity, naturalness, and stability.

This chapter will capture how energy became central to a discourse around economic value during the New Deal period. There exists a long intellectual tradition of appealing to energy as an objective value standard to counter the arbitrary nature of money. This line of thinking, which extends back to the energetics movement and forward to the intellectual programs of Wilhelm Ostwald, Frederick Soddy, and the Technocrats, has been well documented, most notably in the work of Joan Martinez-Alier and Philip Mirowski.[14] Those who called on energy as a replacement for the monetary economy, however, have often been portrayed as isolated thinkers far removed from the mainstream. Technocracy, for example, has rarely been treated as a phenomenon that garnered much popularity beyond the movement itself.[15] The appeal to energy as an alternative value standard, however, circulated widely through American thought during this period. Not only will this chapter demonstrate how energy became central to radical movements, including those surrounding Major Douglas's Social Credit group, but it will also show how it took on a leading role in defining Franklin Roosevelt's social vision for the New Deal.

Once again, how energy sat in relation to value animated this discourse. The engineer and polymath Buckminster Fuller, for example, would critique the crude application of the energy theory of value used by the Technocrats, stressing the importance of human ingenuity in apportioning value to energy, its objectivity found in a unique union between man and the intellect. The libertarian Isabel Paterson would push Fuller's argument one step further, equating energy with money, its objectivity manifest through human desire. The widespread appeal of energy, therefore, transformed it into precisely what it was not: a thing, an ideal form that negated its identity as a social relation little different from money. As a result, energy fell privy to the energy paradox. Just as Erginette adorned the energy archetype too perfectly, the more energy performed the role of energy, the more it came to emulate that which it sought to cancel, the chimera of money.

Technocracy and the Energy Unit of Value

In December 1932 John McCrae, president of the publishing house E. P. Dutton, sent a vicious open letter to the *New York Times* accusing Howard Scott, the spokesman of the Technocracy movement, of plagiarism.[16] Scott's offending article, titled "Technology Smashes the Price System" and published in *Harper's Magazine*, was an extended summary of the philosophy behind Technocracy, explaining the theory of "energy determinants."[17] In the article, Scott challenged the reader to stop and consider, when one eats one's breakfast, the energy source of their bacon and eggs, the newspaper they read, or the automobile that they drive to work.[18] Reflect, he urged them, on the wonder that "everywhere energy, the source, the one and only source of life, is applied to matter."[19] The importance of this call was clear, not least because, as Scott spoke of bacon and eggs, across the nation people were starving. By the time Hoover left office in 1933, more than fifteen million people had lost their jobs.[20] Houses, farms, and businesses had been foreclosed; people went hungry; and itinerant workers took to the streets to find work. Scott blamed this recent human misery on the neglect of one commonsense fact: wealth was energy, not the product of the price system. "We are face to face with a law of nature," he explained. "The law of Conservation of Energy has a perfectly definite social implication," and, he went on, "the only thing that does prevent it is our devotion to a shibboleth—price; and it remains to be seen whether we shall pay for our devotion with our lives."[21]

Illustration of how Technocracy would usher in an era of plenty, with the caption "You'll always have enough to eat." H. I. Phillips, "Technocracy Made Simple: Or, the Technocrat at the Tea Table," *Collier's Weekly*, February 25, 1933, 24.

Scott was not the first to use energy against the shibboleth of price. His entire thesis, McCrae charged, shared an uncanny resemblance to the one put forward by the British radiochemist Frederick Soddy.[22] In his book *Wealth, Virtual Wealth and Debt: The Solution of the Economic Paradox* (1926) Soddy had formulated an energy theory of value. Soddy had first outlined his thesis in a short pamphlet, *Cartesian Economics*, published in 1922. In it Soddy mocked the present farce whereby the monetary economy ran in the opposite direction to the laws of thermodynamics. The financier, little more than a magician, could increase revenue *"ad infinitum,"* ignorant of the fact that *"life is consumption from the cradle to the grave, consumption of that pristine flow of energy we owe to the sun."*[23] With people moving into urban centers, living increasingly artificial existences, the fundamental bond between man and the sun had been lost.[24] By the time *Wealth, Virtual Wealth and Debt* was published, Soddy had recognized that, by forgetting the laws of thermodynamics, society had internalized the structure of debt and was bound to "the fatal destiny, which makes human misery eternal."[25] Money was but a *claim* to wealth, wealth owed in perpetuity. The "fatal destiny" had thus arrived with the naturalization of "Virtual Wealth"—wealth created out of "Absolute Nothing," "the mere fiat of the Human Will."[26] Energy was "Real Wealth," and Soddy argued that "the forms assumed by physical energy are numerous, but all forms are capable of conversion into, and of measurement as heat energy, much as of any account, for the purpose of reckoning, may be transformed into £ sterling, whatever the original currency may be."[27] To cure the recent social ills was a case of reconnecting to "Real Wealth." Scott seized on the importance of this call and offered it as a solution to an ailing nation. By nursing the essential bond between man and the sun, society could once again feast on its bacon.

Unfortunately for Soddy, it had taken the extreme hardship of the Depression for money to unveil its chimeric character to the public.[28] After all, the economist Aaron Director recognized that "in a period of stock-market booms, sex, speakeasies, gang warfare, and other more attractive interests" the new energy creed would hardly have appealed.[29] Even in the desolate postwar years, beyond reform circles, energy barely seemed a viable or necessary alternative to money. When first published in this political climate, Soddy's energy theory of value had only been circulated within a niche audience of monetary reformers and social credit aficionados surrounding A. R. Orage's New Age group and Major C. H. Douglas's Social Credit movement.[30] Then, as periodicals devoted more articles to Technocracy than to the presidential candidate Franklin D. Roosevelt, Soddy's message (through Technocracy) had, for a brief moment, broken through.[31]

Contrary to McCrae's plagiarism claim, however, Scott had been developing his own theory of "energy determinants" for almost a decade. Scott had first posited his theory of "energy determinants" in 1919 while associated with the Technical Alliance, formed around an aging Thorstein Veblen at the New School of Social Research in New York. This group responded to a series of articles published by Veblen in 1919 in the *Dial* (later turned into a book titled *The Engineers and the Price System* [1921]). In these articles, Veblen saw the entire price structure propped up by the "conscious withdrawal of efficiency."[32] To counter this tension, the Technical Alliance would return the system to those who could manage it— the engineers. At the time, Scott proposed an energy-based production system that would eradicate the profit motive, and for the next decade he could be found waxing lyrical about "energy determinants" in Greenwich Village taverns, while the nation enjoyed the bounty of the late twenties.[33] Eventually Scott would found the Committee on Technocracy along with Walter Rautenstrauch at Columbia University, where, with the assistance of around forty engineers and technicians, they documented the industrial and agricultural outputs of the United States over the past hundred years, advocating for the organization of society along technical lines. In August 1932, when Rautenstrauch announced the results of the Energy Survey of North America, its findings spoke to a nation in distress.

Money as a Rubber Yardstick

How ridiculous would it be to build a house with a yardstick made of rubber? Well, Scott explained, an implement as ludicrous as a "rubber yardstick" was currently measuring modern civilization.[34] This instrument was the dollar. Before the First World War, money preserved an illusion of hard rule, with an aura of credibility and confidence invested in the gold standard by the public.[35] This was shattered after the war, following a decade of financial instability. German hyperinflation persisted from 1921 to 1924, discrediting the supposed "hardness" of money. In 1929 the Wall Street Crash revealed money to be entirely elastic. From 1927 onward, the "orgy of mad speculation" forced the stock market to lose any illusion of value.[36] As business activity subsided, prices soared with little discernible relationship to value.[37] On October 23, 1929, the speculative bubble burst and revealed the financial system to be an illusion.[38] It was little more than collective will. Looking back at this moment, the economist Irving Fisher described the struggle that his readers had in getting used to the fact that "all his life, he has been the victim of the 'money illusion.' This illusion is the notion that the monetary unit is always the same, so that it can serve as a measure of value of other things, but does not need to be measured itself."[39]

With the dollar exposed as a "rubber yardstick," energy came to embody the stability money lacked. Energy, Scott pointed out, was a metric system that measured nature through nature, wealth through wealth, rather than being controlled by private interests. It had "intrinsic" rather than "extrinsic" value. The premise that energy was an intrinsic value, Joan Martinez-Alier has documented, sat within a long tradition of reform economics.[40] Soddy himself recognized this, noting publicly in an article in the British *Daily Mail* that the energy theory proposed by the Technocrats was derivative of his work and that the "substitution of payments in 'energy certificates' has certainly been advocated by every form of reformer across the channel."[41] For instance, Patrick Geddes had discussed the search for an intrinsic value by drawing on John Ruskin's critique of capitalist value production. In his book *John Ruskin, Economist* (1884), he disputed that there was no such thing as "intrinsic value" and asked that we "walk out into the world, look about us, try to express loaf and diamond from the objective side in terms of actual fact."[42] Quoting Ruskin's determination that "a sheaf of wheat of given quality and weight has in it a measurable power of sustaining the substance of the body; a cubic foot of pure air, a fixed power of sustaining its warmth; and a cluster of flowers of given beauty, a fixed power of enlivening or animating the senses and heart," Geddes made a claim for intrinsic value.[43] While Geddes posited intrinsic value in relation to other physiological, emotional, ethical, and aesthetic standards, Scott and Soddy limited intrinsic value to the physical unit of energy.[44]

With the dollar being an arbitrary measure prone to pop overnight, the energy unit appealed because of its stability.[45] Unable to be transformed into a simulacrum, energy connected value to an intrinsic standard in the physical world.[46] Scott maintained this fact by arguing, "A dollar may be worth—in buying power— so much to-day and more or less to-morrow, but a unit of work or heat is the same in 1900, 1929, 1933, or the year 2000."[47] Value, when taken to be a "force of desire," had no metrical equivalent and could be manipulated by subjective whim. On the other hand, Scott asserted that "a pound of coal is always a pound of coal" no matter who controls the system.[48] With the recent hardship accredited to a faulty and pernicious price system, Technocracy would eliminate subjective standards that could not claim authority as intrinsic wealth. In the ideal Technocratic State energy abolished the distortion of value, with the cost of goods measured against a "balanced load" period (correlate to the production limit of any cycle). Cost would correspond to the energy and raw materials used in producing the article.[49] No longer would bacon and eggs be priced according to scarcity values, but their cost would match the quantity of ergs spent in production. Money in the Technocratic State would be replaced by "energy certificates," with an item's value determined by the number of energy units needed to produce and distribute it.

By March 1933, a rift between Scott and Rautenstrauch signaled the end of Technocracy as a movement, with Rautenstrauch distancing himself by establishing the Continental Committee on Technocracy. Critics attacked the energy theory of value, condemning the group for a faulty understanding of economics and accusing them of manipulating results. They also dwelt on the reasons why energy had caught hold in the public imagination to such a degree. In one analysis, the economist Aaron Director attributed energy's appeal to the simple fact that the "natural sciences have great prestige."[50] As he explained, although "popular belief in their doctrines varies inversely with popular comprehension," the use of phrases such as "'energy determinants,' 'conservation of energy,' 'decision arrivation,' and the like is irresistible and leaves us aglow with knowledge and erudition."[51] In contrast to the recent spate of garbled economic, political, and philosophical models that had failed one by one, the erg benefited from clarity and its authority as physical science.

With the erg legitimized as hard science, the writer and sworn Technocrat Harold Loeb questioned the necessity of a voting democracy swayed by opinion when there was a "yardstick" in energy.[52] Personal opinion, individual politicking, and private interests, he noted, had led to rioting, mass panic, and destitution, while the erg had been infallible so far. Despite the threat to liberal democracy posed by energy's autocratic rule, Herbert Hoover's attack on the Bonus Army earlier that year in Washington, DC, had left democracy appearing little better than a sham. The rubber politics exhibited by Hoover's tenure would be hard to defend against an energy "yardstick" that accrued political authority from science. The year Roosevelt entered office, nations across Europe were rejecting liberal democracy and embracing dictatorship.[53] Dismissing parliamentary democracy as limp and ineffectual, these dictatorships promoted a new anti-liberal outlook.[54] The disillusionment over the lethargy of liberal democracy was not confined to Europe; on the eve of Roosevelt's election, voices expressed the need for a similar model of hard rule in America.[55] With the precarious circumstances at hand, many feared that the United States had to implement extraordinary circumstances, a form of dictatorial rule that would undermine the regulatory processes promised by the Constitution.[56] The efficiency and decisiveness shown by Mussolini in Italy and Stalin in Russia had set an example, and Walter Lippmann memorably begged Roosevelt: "The situation is critical, Franklin. You may have no alternative but to assume dictatorial power."[57] While some believed dictatorship to be preferable to endless debate, others less susceptible to the cult of personality recognized energy to be a more impartial rule than the political ego. As a neutral "yardstick," energy was free from the egomania and subjective whim expressed by leaders such as Hitler and Mussolini. With personality increasingly saturating politics, energy's stable and, more importantly, neutral temperament appeared a worthy attribute. Energy would not be subject to bribery, prejudice, weakness, greed, or any other human deficit.

Others, however, recognized the appeal of this energy despot to be nihilistic passivity. It was, one commentator suggested, "a masochistic surrender to destruction" and a release from the "barrenness of the system by which they live today."[58] As many observers pointed out, people turned to energy out of sheer desperation. This was the conclusion of the *Brooklyn Daily Eagle*, whose journalists had taken to the streets to find out what the local community thought of Technocracy. In doing so, they reported the opinion of a female typist who confessed that, as she currently had absolutely nothing to lose, she might as well try Technocracy: "if it is true," she mused, that "they are going to do away with money, then everybody else would be as bad off as I am."[59]

Although the turn to energy reflected a growing pessimism in the current order, others believed, more optimistically, that energy could reinvigorate ethics and provide a new moral standard beyond money. The writer Harold Loeb, founder and editor of the avant-garde journal *Broom*, which from 1921 to 1924 had been a central forum to celebrate the aesthetics of the machine age, trusted that it would allow for a new flourishing of creative spirit.[60] In his book *Life in a Technocracy* (1933) Loeb evoked a world in which religion, art, literature, love, and government were released from the dictates of the market.[61] Religion, he explained, had been replaced by the "Mysticism of Money," and now, he hoped, energy could replace money as an aesthetic and ethical standard.[62] Similarly, "human energy," when freed from commercialism, could be "applied to finding ways of life that satisfy, to creating values which uplift the spirit," and thus transform both man's "inner psychic sphere" and his outer physical existence.[63] In this utopia, "buildings would no longer rise in competitive ostentation; advertisements would no longer entice the unwary to increase his wants; jazz would change its tempo; and the prose, the movies, the theater would find other stories to tell, other ideals to vaunt."[64]

While Loeb hoped that energy would provide a more liberating yardstick than money, the poet Archibald MacLeish took this to be a naive position. He pointed out that energy went "no farther on the philosophic side than to offer to an increasingly childish humanity another mechanical nurse, another external authority, and another absolute pattern."[65] Far from granting new freedoms, the logical conclusion of this position would be a "thermodynamic dictatorship of the spirit."[66] After all, MacLeish explained, energy would be no less oppressive when it was imposed as an absolute standard than the excessive commodification exerted by Mammon. The eagerness, however, to replace one absolute standard (money) with another (energy) highlighted another part of energy's appeal: it provided an absolute standard in a world disintegrating into chaos. This point was recognized by the economist and literary critic Henry Hazlitt. Hazlitt described how the attack on absolute standards—from figures such as Charles Darwin, William James, Albert Einstein, Sigmund Freud, and Friedrich Nietzsche—had left energy as the only absolute left in a world disintegrating into relativism.[67] Being, however, one of the strongest defenders of relativism and a vocal advocate for the marketplace, Hazlitt had to remind Scott that "energy had no meaning except in terms of value."[68] It was as useless, Hazlitt concluded, to base a currency on energy as it was to attach it "to Pi—on the ground that this never varied, but was always 3.1416—and that it be made redeemable in diameters, because there is an infinite number of them in every circle."[69]

While many praised energy as the only absolute left in a world fragmenting into relativist chaos, some female critics suspected that the new appeal to energy was part of a patriarchal conspiracy to dominate women. The efficiency expert and home economist Christine Frederick compared the energy models posed by the Technocrats to "the ambitions and cock-sure assertions of boys playing with mechanics in their workshops."[70] To her, it neglected the reality not only of being a woman but also, more importantly, of being a consumer. Defending Mrs. Consumer, Frederick wanted to know how her designer dressmaker was going to be paid if everything was to be reduced to erg rations. She asked, "If my exceedingly artistic dressmaker is going to be obliged under Technocracy, to give her dresses away to all who present enough energy certificates to cover the bare cost of energy used up in making the dresses, then it is clear as crystal to me that a large squad of policeman will be necessary to guard Madame B's place of business. It will be mobbed by women who will be out for the greatest bargain of the age!"[71] Under a regime dominated by energy, Frederick believed, "women would suffer perfectly ghastly limitations."[72] This would involve an attack on her fashion, her leisure, and even her personal relationships. With the male identifying himself with the "Great God Calorie," energy would impose a masculine order that would crush the very essence of femininity, desire. It was, Frederick believed, important to note that woman "long, long ago ceased to buy her clothes and a great many other things on the basis of the value of the raw materials or the energy consumed."[73] Being a female, Frederick concluded (drawing on popular gender norms), "I am more interested in what merchandise does to people than I am in the merchandise itself, or the foot-pounds that went into it, or the machinery that made it."[74] While femininity was aligned with consumption, energy, on the other hand—and the positive values that went with it, such as objectivity, neutrality, and intrinsicality—was characterized as male. This was in opposition to Erginette, who, like Mrs. Consumer, inflated value ad infinitum.[75] To counter this blatant attack on women by energy, Frederick wanted to remind the public that men do not "live by energy alone, *nor by use alone.*"[76]

Although opinion was divided about the merits and shortcomings of Technocracy, energy fulfilled a range of cultural needs at a moment when the monetary illusion had been revealed to be little better than a specter.

Solar Credit, Anti-Semitism, and Usury

The Depression had exposed the private interests, subjective whims, individual egos, and political fancies in control of the price system, and as a result, energy, for many, was a tempting alternative to money. Energy, after all, supported hardness over rubber, intrinsic value over extrinsic, absolute standards over relative, and objectivity over subjectivity. Technocracy was not the only reform movement to turn to energy for these qualities, but during this period, both in the United States and in the United Kingdom, other underconsumptionists, such as those involved in the Social Credit movement, turned to energy to bolster their programs.

Originating from the writing of Major Douglas, Social Credit's moment came (like that of the Technocrats) during the 1932 economic slump.[77] During this period, a Social Credit government was founded in Alberta, Canada, and in the United Kingdom the Green Shirts became a formalized movement spreading Major Douglas's gospel.[78] The origins of Social Credit emerged when Douglas, at the time an engineer in a factory, noticed that the goods produced were always of a higher value than the wages going to laborers and shareholders, leading to an inevitable deficit in purchasing power. To describe this process, Douglas posed his well-known A+B theorem, which operated on the proviso that individual wages (A) were always lower than needed to buy a product when the costs of raw materials and bank cost were added (A+B). The solution posed by Social Credit was to provide a subsidy (in the value of B) to the consumer, to match purchasing power with the goods in the economy. To the militant leader of the Green Shirts John Hargrave, however, the A+B theorem could be put a different way: "Money Power: *Get out of the way!—And allow solar-energy (real wealth) to reach us.*"[79] Emerging from Hargrave's youth group, the Kindred of the Kibbo Kift, the Green Shirts drew on military strategies to promote Douglas's Social Credit philosophy.[80] Hargrave, however, would later describe the entire movement as a battle between energy and money. In his memoir "They Can't Kill the Sun," Hargrave reduced Social Credit to a "solar vision," an "Elemental Struggle—the war of Light against Darkness," of energy against money, "breaking from the psychological and spiritual sphere, first into socio-economic conflict, and finally into direct physical (military) war."[81]

Breaking down the differences between Left, Right, and Center as defined across the political landscape, Hargrave recognized that purchasing power was a contract between man and the sun, the A+B theorem being *"the key-formula of the economics of solar-energy."*[82] Because solar energy pouring down on the earth was *"free of charge,"* Hargrave rejected the well-known dictum that you "cannot have something for nothing." True, Hargrave admitted, "you cannot have something from nothing," but he added that the sun is not nothing and, as a result, every item on the market was in essence solar energy. In this sense, although Scott called for the erg to be attached to a value factor (in relation to the cost of production), Hargrave interpreted this as just another "financial game of let's pretend."[83] Instead, Hargrave's helio-economy was grounded on the earth as a "lump of sun-material" supported by unlimited free energy. The cost accountancy model had to reflect the actualities of energy, and, Hargrave argued, it was futile to attempt to pay for a few "ergs" of solar energy with the coins in his pocket, even if he offered a check reading *"To Old Sol, or Order!"*[84] In this way Hargrave conflated Social Credit with energy, describing it as "the Gospel of Common Sense."[85] In turning away from energy, all political movements had collapsed into politicking, partisanship, and subjectivity. To counter this, Hargrave announced that "a new political dynamo has already emerged—*the recognition of Solar-Energy as the source of all Life and Wealth.*"[86]

Hargrave thus echoed Scott and Soddy in aligning his political movement with energy. For both groups, there was no moderate position or party line; it was one or the other—money or energy. Pushing this to the extreme, Scott discredited rival reform movements (including Social Credit) for failing to recognize Technocracy as anything but pure fact. Scott, for example, refused to accept Upton Sinclair's claim that his End Poverty in California movement was "merely a practical technic for bringing Technocracy to pass."[87] Sinclair, Hargrave believed, had mistaken Technocracy to be an interested political party, rather than a movement grounded in material fact.[88] Other reform groups that self-identified as heirs of Technocracy, Scott asserted, were similarly in the wrong. Claims of affiliation made by the Townsend Plan and Premier William Aberhart's Social Credit Party in Alberta also revealed how little they valued the unbending truth energy presupposed. All of these political parties, Scott felt, held "the quaint belief regarding the omniscience of multiplied human opinion, namely, that if human opinion is expressed in large enough numbers it can alter the laws of gravitation."[89] Instead, Scott determined, energy negated in-betweens, and as a result, there was no room for "multiplied human opinion." Just as Scott accused the Social Credit movement of falling back into party politicking, Hargrave lodged a similar attack against Scott and his Technocratic movement. In his "Solar Vision" the A+B theorem had been revealed as the *"key-formula in the economics of solar-energy,"* not in Technocracy's theory of "energy determinants."[90]

Not only did radical reform parties turn to energy to justify their programs, but the Far Right used energy to fuel anti-Semitism, adopting Soddy's energy theory of value as justification. With the rise of fascism abroad and growing paranoia about the economy, anti-Semitic sentiment increased during the New Deal period.[91] As hate rhetoric intensified, the Jew (represented by the international financier) became a popular scapegoat for the financial collapse.

The rhetoric that made energy conducive to this discourse emerged out of long-held conspiracy theories that connected Jews to money. Brooks Adams had drawn on this symbolism in his "Law," characterizing the international financier as the hoarder of energy, the final stage before civilization's collapse.[92] This stereotype extended back to a period when Jews were forced into financial occupations such as money lending, tax collection, and banking, which depended on money lent at interest. With money and the Jew framed as synonymous, anti-Semitism utilized a correlate body of adjectives: the Jew was unnatural, dishonest, virtual, devilish, a floating referent that suffered from endogenous excess.[93]

This analogy was repeated in discussions of Social Credit, with Arthur Kitson and Major Douglas tying their economic thought to this imagery. However, the association between energy and anti-Semitism was best articulated by Soddy, who, in his early pamphlet *Cartesian Economics*, had employed the laws of thermodynamics as an attack on the "usurers of the world," who, like Shylock, "are bent on a pound of flesh next to the heart."[94] By 1939 this connection became increasingly explicit, and a caricature of a fat Jewish businessman would adorn a pamphlet written by Soddy with Walter Crick, *Abolish Private Money, or Drown in Debt*. This pamphlet, which quoted from the famous anti-Semitic forgery *The Protocols of the Elders of Zion*, pinpointed the transition from "intrinsic wealth" to legalized usury at the moment Cromwell let Jews settle in England.[95] By making this differentiation, Soddy equated "Virtual Wealth" with the Jew and "Real Wealth" (bound by the laws of thermodynamics) with the anti-Semite.

Where anti-Semitism was implicit in Soddy's economic theory, the anti-Semitic radio evangelist Father Coughlin, the so-called father of "hate radio," made this association most forcibly. Every Sunday Coughlin preached across the airwaves to an audience of ten million.[96] Fueled and inspired by Soddy's writing, Coughlin rallied support for an end to monetary manipulation.[97] With his parish in the Shrine of Little Flower in Royal Oak, Michigan, near Detroit, a city badly affected by the Depression, Coughlin combined religious doctrine with his disgust at the recent economic crisis.[98] Coughlin told his captive audience that gold was emulating God, by defying the limits of wealth and failing to accord with the laws of thermodynamics.[99] Only God, who was not bound by physical laws, was permitted to extend into the infinite, Coughlin explained. In spite of this, money, as "Virtual Wealth," grew beyond physical limitations, thus imitating God—guilty, therefore, of the worst act of blasphemy. Coughlin, therefore, read Soddy's analysis through Leviticus: "material wealth is not infinite in duration" but perishes.[100] By 1933, Coughlin's politics had become more extreme, and he blamed both the First World War and the Depression on the manipulation of international finance controlled by the Jews. Originally an avid supporter of the New Deal, by 1934 Coughlin had turned his back on Roosevelt, fueling a conspiracy about Roosevelt's Jewish roots and establishing a new political party, the National Union for Social Justice. Opening his later book *Money! Questions and Answers* (1936) with the challenge "your money or the Christian concept of life," Coughlin set up a dichotomy. On one side are artificial laws manipulated by international finance, and on the other are the laws of nature.[101] Money (represented by the Jew) was founded on the principle that one could extract "something from nothing": it was *contra naturam*—in defiance of the laws of thermodynamics. In opposition, the anti-Semite drew his moral authority from energy—in line with both religious and natural law.

The poet Ezra Pound extended this principle into both a literary style and a political dogma, charactering any defiance of this moral energetics "Jewsury."[102] A fan of Coughlin (as well as Brooks Adams) and a Social Credit aficionado, Pound sought to reconnect both the monetary and the referential sign with their material basis.[103] In fact, Paul Morrison has argued that Pound's entire poetics "might be characterized as an attempt to recover the priority of use over exchange value for the modern world."[104] Pound had been initiated into the world of monetary reform through the Social Creditors surrounding the New Age group.[105] He became fanatical about economic reform during the Great Depression and increasingly disillusioned with Roosevelt's New Deal, believing that it was too indebted to private finance. During this period, Pound became a devotee of Mussolini and the German economist Silvio Gesell. Gesell's "stamped money," known as "rusting bank notes" (implemented in kind for a brief period in 1932 in the town of Wörgl, Germany), would be a self-taxing currency that expired after a month. This ensured that money functioned like energy, being self-destructive—the banknote could not be hoarded as it gained interest.[106]

Pound also applied his economic thinking to a theory of language. While the appeal of Gesell's script for Pound was that, like the energy certificate, it eliminated compound interest, preventing the dissociation of the sign and its referent, Pound's equivalent—the ideogram—collapsed this distinction and turned language into a material embodiment of energy. For Pound the continued appeal of the Chinese ideogram (famously articulated by Ernest Fenollosa) was its energetic quality that mirrored the transfer of force inherent in all natural processes.[107] Whereas the boundary between thing and action was inherent to Western syntax and "monetary form," for Pound Chinese words revealed "nature as a vast storehouse of forces."[108] The futility of "picking up a man and telling him that he is a noun, a dead thing rather than a bundle of functions!" became as much an energetic proposition as a political or poetic guideline.[109] For Pound, the "Jew" represented a self-referential "thing" that pivoted like the free-floating signifier, whereas Pound called for a referential framework that tied both semantics and society to a natural energetic. Energy, therefore, provided a radical critique of the contemporary economic and (in the case of Pound) representational order, which, in the hands of fascist sympathizers, could be expanded into an anti-Semitic critique.

The New Deal's "Energy Yardstick"

The rhetorical promise posed by energy was not limited to radical reform movements and anti-Semitic rhetoric. Franklin D. Roosevelt, on the campaign trail when Technocracy was at the height of its popularity, also drew on the symbolism of energy as a campaign weapon. For example, at a high-profile campaign event in Portland, Oregon, in September 1932, Roosevelt drew on the image of the "yardstick" to drive home his political message. Condemning the recent crash of Samuel Insull's holding company, the crowd cheered as Roosevelt promised the development of a "yardstick" to act as a government- and community-sanctioned minimum price for energy.[110] It would be, he suggested, a "'birch rod' in the cupboard to be taken out and used only when the 'child' gets beyond the point where a mere scolding does no good."[111]

The "yardstick" slogan had not been invented by Technocracy or Roosevelt but was a popular mantra dating back to reform movements emerging at the turn of the century, and it had since circulated through progressive rhetoric to denote government regulation.[112] It had, however, been used frequently in discussions of Technocracy—the "energy determinant" described evocatively as a new social "yardstick" for society.[113] Some commentators, such as Wayne W. Parrish, would even confuse the energy determinant with the "yardstick" to evoke an "energy yardstick that transcends all political doctrines."[114] Seen alongside the Technocrat's "energy determinant," the "yardstick" metaphor evoked a similar symbolism, promising a fairer standard of living founded on real, rather than virtual, wealth.[115] Drawing on this symbolism, Roosevelt made a similar pledge to the people, describing how energy belonged to the public and declaring that "no commission—not the Legislature itself—has any right to give, for any consideration whatever, a single potential kilowatt in virtual perpetuity to any person or corporation whatsoever."[116]

Roosevelt held to his promise, and one of the central directives of the Tennessee Valley Authority (TVA) was the provision of an energy "yardstick," a government-backed wholesale price for energy to prevent private extortion. To solidify the TVA's public image, the "yardstick" metaphor was cleverly manipulated by the administration, which utilized its symbolism to move beyond a functional discussion of price, to transform it into a social ideal.[117] One journalist in the *Literary Digest*, for example, witnessed how the "vision and idealism seem to light up his announcement of rates, so that even rate kilowatt-hours, as reported in the press, evoke a certain romantic thrill."[118] With the first rates released by the TVA pointing to the disparity between a Baltimore consumer paying $2.63 for 50 kilowatt-hours, the New Yorker $3.05, and the St. Louisan $2.05, the TVA rate, which started at $1.55, became a basic energy rate that echoed the Technocrats' intrinsic energy determinant.[119] Like the erg unit, the "yardstick" backed by the government would be immune to private interests and personal racketeering. Although valued at $1.55, a signifier that tied energy immediately back to an economic value, the energy "yardstick" offered a semantic promise: the government rate would stand as the "objective" cost of energy. Whereas the utilities would be founded on rubber, the federal government under Roosevelt's leadership would be based in energy.

By conflating state-backed rates with an intrinsic energy determinant, Roosevelt naturalized the extension of government. And yet, it did not take long before the "energy yardstick" was revealed to be elastic.[120] One commentator recalled Einstein's formulation: "a yard is a little more than a yard or a little less than a yard depending on which way it is going."[121] In spite of the supposed legitimacy of this "yardstick," he added, "It all depends on whether you are in the old-fashioned business of making and selling electricity or whether you are out to lick the Power Trust. If you are a follower of Senator Norris, a measuring rod is still a measuring rod even when it is pickled in subsidy."[122] To the private utilities the energy "yardstick" was no less "elastic than rubber." By April 1935, with charges of extortion and mismanagement on the ledger, David E. Lilienthal would be in front of the House Commerce Committee defending the "yardstick" against the accusation made by the utility companies that it was no better than "rubber."[123]

From its inception the symbolism of the New Deal had been steeped in the language of energy. In 1932, before Roosevelt had even coined the term, Stuart Chase argued that the United States needed "A New Deal"—"a new philosophy of energy applied in space through time."[124] Chase had a long involvement with Scott, being part of the original Technical Alliance. By Roosevelt's election, Chase was America's most widely read economist and propagandist for the New Deal.[125] Throughout this period, Chase advocated a planned economy "typical of the Russian experiment" that would return America to its energy basis.[126] Trained as an accountant, Chase joined the Federal Trade Commission in 1917 to investigate the meatpacking industry. During this period, he was charged, along with his colleagues, with having "communist sympathies"; the accusations against him even extended to him carrying a red flag at an anarchist parade in Chicago and having juggled the "packer's profits" so that the "blackest case may be made against them."[127] Although cleared of these accusations, Chase remained an avid supporter of the communist revolution.[128]

Chase's book *A New Deal* (1932) put forward a vision of a "third road for the Nation," which later became associated with Roosevelt's radical program. Chase pointed out that "neither mysticism, political rhetoric, nor contemplation of the navel will get kilowatts out of Niagara."[129] Instead, the country needed to reconnect with its "energy basis." Drawing heavily on Soddy's book *Wealth, Virtual Wealth and Debt*, Chase argued that the nation was suffering from the divide between "Real Wealth" and "Virtual Wealth," which had latterly been ingrained "in Western mentality."[130] This cleft could be "blasted by facts, not by reason or rhetoric."[131] The 1929 Wall Street Crash was one such blast. What was needed now were "facts, *facts*!"—an "ear-splitting, blinding, stupefying word, 'facts.'"[132] For Chase, this knowledge would come from working "for a time with the physical units of tons and kilowatts and man hours rather than in the metaphysical units of money and credit and price!"[133] Money was a "metaphysical unit," he argued, whereas a fact was "an occurrence in space and time which, if seen by one competent observer, is capable of being verified by another."[134] Those "physical units of tons and kilowatts and man-hours"—that energy was the "clear trail" the "fact" to be picked apart "nugget by nugget."[135]

In his book *The Economy of Abundance* (1934) Chase set out to trace the robber baron depriving society of its energy. To do so, he drew on E. W. Zimmermann's concept of an "energy basis" to imagine "the foundation of civilization."[136] This "energy basis," he maintained, "determines the choice of materials which can be utilized; it sets a definite limit to the size of performance; it governs the degree of mobility, and in general controls the arts, and through them shapes the institutions—material and non-material."[137] Looking at the current conditions of America, it was clear "what direction the vanished energy has taken."[138] Echoing Veblen, Chase described how obsolescent wastage could be found in all stages of the industrial process, such as "Wall Street 'extras,'" overproduction, "foreign gifts," and "pleasure motoring."[139] Despite New York possessing more energy than the entirety of Europe in AD 1400, this energy was not translated into wealth in terms of "human use and enjoyment." Instead, Chase resolved, it was "largely misplaced energy crystallized in stone and steel."[140] Even though there was an abundance of energy, Chase concluded, America was just as poor in 1830 as it was today in terms of "human calculus."[141]

To demonstrate this, Chase juxtaposed the "energy basis" of North America to handcraft economies. In an earlier study, *Mexico: A Study of Two Americas* (1932; illustrated by the Mexican radical Diego Rivera), Chase compared the energy basis of North America with that of postrevolutionary Mexico. He did this by comparing the life of Middletown (the typical American town captured by Robert and Helen Merrell Lynd in their famous study of Muncie) with that of Tepoztlán, a Mexican town of a similar scale.[142] Throughout the study, Chase described the many ways in which Tepoztlán was connected to its energy base, unlike Middletown, which, powered by "our billion horsepower," had run off course.[143] Although Middletown had clear advantages in terms of education, health, and modern technology, it had no conception of "basic biological and psychological needs."[144] Tepoztlán, in contrast, had "never lost these values. It works, plays, worships, attires itself, composes its dwellings in the normal rhythm of *homo sapiens* upon this planet, without abnormal effort, without waste. It knows what life is for because every move it makes contributes to a legitimate function of living."[145] Not only, therefore, had America exceeded the limits of its environment, but Chase argued that it had defied "human nature."[146] "Surfeited with an undigested mass of functionless material," the United States had become "uncomfortable," "constipated," "compressed," "weary," "puzzled," and "perplexed."[147]

When the paradigmatic New Deal town of Norris Village was settled four miles from the Norris Dam, it resembled Tepoztlán rather than Muncie.[148] This was hardly surprising, given the involvement of the Regional Planning Association of America (RPAA) in inspiring the planning vision, even if they had been excluded from consultations on the plan.[149] Interest in handicraft economies came from the top, with first lady Eleanor Roosevelt, Lou Henry Hoover, and Grace Coolidge working together to promote the movement.[150] Despite its proximity to the largest hydroelectric dam in the region, Norris Village acted as a potent symbol: it would represent technological modernity while maintaining the impression of being a model handcraft society grounded in energy—in the rhythms of nature, rather than in those of the capitalist megamachine. Energy, therefore, provided a potent symbol for the New Deal of a grounded economy, backed by the government rather than the price system.

Critiques of the Energy Theory of Value

By 1933 "electricity dollars" were being slipped into circulation and operating like money. The *New York Times* reported the story of Valeska Becker, a student at Antioch College who, with no money to spare, had offered up "twenty kilowatts of my energy in working for humanism" for the collection tray.[151] Across the country people debated the significance of Technocracy for Christianity, raising the question of how far energy could get you in the economics of salvation.[152] Would 20 kilowatts of energy be enough to grant eternal redemption, or provide the energy needed to ascend a step up the ladder? Could it be converted into an equivalent measure of grace? As salvation could not be reduced to energy units, in divine matters energy would be forced to operate as money, a correlate of subjective value—the desire for redemption.

That the "energy certificate" suffered from the same value fluctuation as money was not lost on Technocracy's many critics. Not only had John Erskine captured this problem in his satire "Erginette," but other critics also rejoiced in pointing out the many fallacies inherent in the energy theory of value. Barton Griffiss, head of the economics division of the Carnegie Institute, pointed out that Scott's theory of "energy determinants" failed to give proper consideration to the "human element" in economic transactions.[153] It reminded him too much of the "old labor theory of value" that failed to take into account that the energy unit put forward by a skilled laborer differed in value to that of an unskilled laborer. Pointing toward the impossibility of quantifying the energy generated by the lawyer, the doctor, or the author, he questioned how one was going to quantify energy when the amount of energy spent did not reflect the amount of work done—let alone the use value of such work.[154] Recognizing a similar problem, one Democratic representative had to remind the House that there was no objective way to assign an energy determinant to every category of work. "Scientists have struck a snag," he explained, "because they cannot estimate the horsepower used up in gum-chewing Congressional speeches, Marathon dances and research work. It is figured that it will take another 1000 years to get a fair base for the domestic allotment of energy-certificates."[155]

How to calculate the energy used in different forms of labor, therefore, was not as easy to quantify as first assumed. Not only did this difficulty arise at the production end, but a similar challenge was also faced as people sought to calculate the value of energy's end use—its social worth. Echoing William James, Walter Lippmann reminded Scott that energy was not equitable with social good. Pushing Scott's thesis to its logical conclusion, he argued that "if Mr Scott's assumption were true, the same number of calories consumed in shooting off a cannon and building a house would produce the same social results."[156] The human, Lippmann added, was "a living soul" who could direct energy in any way so desired; furthermore, a "human society, being composed of millions of living souls, is the most complex of all phenomena which men attempt to study."[157]

To Technocracy's many critics, it was evident that energy could not escape the distortion of value. In fact, the economist Henry Hazlitt had to conclude that energy could only be recognized within the industrial system, when it was converted back into economic value. He explained:

> The energy that pours over Niagara Falls is the same, harnessed or unharnessed, but what concerns us is whether it is harnessed, and for what purposes. Man cannot create energy; he can merely transform it. What he attempts to do is to transform it into forms of higher and higher value. That being so, it is the value that matters, not the amount of the energy. If it were only energy that was important, we should all want to live on the sun, which discharges 3.79×10^{33} ergs every second, of which all the planets taken together get only 1/120,000,000 part.[158]

If energy, as Hazlitt recognized, could only operate within industrial society as value, another issue that had to be confronted was the regulation of supply and demand. One economist aired his concerns about the emergence of a bootleg economy creating a black market to distort the value of "energy money." He imagined a situation in which the hills might be full of "illicit stills" that could produce and transmit energy secretly to towns and cities.[159] He even feared that the transmission of power by radio could be invented, causing a large surfeit of energy on the market. All these factors made "energy extremely unstable" as a value unit.[160] Not only did this highlight the fact that energy would be just as susceptible to value fluctuation as money, but it raised the question, what did energy have that money lacked? George Soule, editor of the *New Republic*, posed this problem, pointing out that if the nation under Technocracy managed to have such heightened control over its energy, why couldn't old-fashioned money remain an equally effective value standard?[161] The central issue, Soule recognized, was "to regulate production, investment, credit, prices and incomes," not the need to invent a new currency itself.[162] This wave of criticism anticipated the later attack on rationalistic planning by the free market liberal Friedrich Hayek, who drove home the fact that energy, despite claiming intrinsic value, was always just as unstable as money. Hayek would take aim at the "energy theory of value" held by Soddy and the Technocrats, deriding its false "objectivism" and faith that the world could be entirely known and reduced to one standard. Hayek mocked the absurdity that anyone would "treat the various things not according to the concrete usefulness they possess for the purposes for which he knows how to use them, but as the interchangeable units of abstract energy which they 'really' are."[163]

Buckminster Fuller's Rubber Yardstick Made of Energy

Although market liberals, such as Hazlitt and Hayek, were adamant that all standards (including energy) were dependent on subjective desire, others clung to an intermediate point, locating energy in between subjectivity and objectivity. Such a message was best epitomized by the designer R. Buckminster Fuller, who added his own spin on the "yardstick debate," envisioning a rubber yardstick made of energy. Being too liberal in spirit to accept autocratic rule by energy, Fuller developed an inclusive spatial model that returned economic, social, and spatial structures to energy. However, in the same move Fuller pushed energy's objectivity to its limits, grounding it in the logic of rubber. Hanging out with Scott in his favorite Greenwich Village haunt, the café Romany Marie's, Fuller absorbed the dogma of "energy determinants" but formulated an explicitly teleological energetics.[164] His method of Dymaxion design would be at the heart of this project, which involved a form of modeling on a four-dimensional plane, allowing for the manipulation of energy patterns for human advancement.

In his book *Nine Chains to the Moon* (1938), Fuller pointed out that the recent fatigue besetting man was down to the presupposition "that this surface alone is reality," in spite of the fact that human beings were "nurtured and energized by the ultraviolet and gamma rays, as well as by most powerfully penetrating, highly energized cosmic rays."[165] Words, things, monuments, and houses were antiquated thought structures superimposed on top of energy, rather than grounded in its logic. Although man was "a self-balancing, 28-jointed adapter-base biped; an electro-chemical reduction-plant, integral with segregated stowages of special energy extracts in storage batteries, for subsequent actualisation of thousands of hydraulic and pneumatic pumps," the human environment, along with models of thought, relied on several degrees of abstraction manufactured through idealistic systems controlled by the price system.[166] Fuller had personal experience of the damage wrought by this fissure. Not only had his family been subject to periods of abject poverty, but he also blamed this abstraction for the death of his young daughter from influenza in 1922. This tragedy, he maintained, was caused by the failure of the health system, which depended on entirely mismatched systems of "patterns arbitrarily selected" and based on the necessity "to make money first."[167] To fix this, Fuller drew on his training in the US Navy to develop a form of "energetic geometry" that relied on augmenting energy patterns.[168]

Like the ideas of his contemporaries, Fuller's energetic geometry would appeal to the neutrality promised by energy. Echoing the argument that energy demonstrated a "'non-political'—political logic," Fuller was adamant that his "energetic geometry" was "utterly independent of any need for authority beyond that to-self-by-self for initiation of its study and development."[169] It instigated *"'the integrated, teleologic objectivity of the full gamut of the exact sciences,'* no more–no less."[170] Fuller's "teleologic objectivity" thus returned energy to its a priori logic, and yet, as it was improved through human systems, it was returned to an episteme in human conditioning.

Fuller recognized, for instance, that energy had the fastest capacity to accrue value.[171] This sense, he argued, was reflected in the government's recent departure from the gold standard (which retailed at a "paltry price" of $418 per pound of gold) to energy (the photon retailing at approximately "1 ½ billion dollars per lb" in finance capital).[172] This bifurcation between a basic energy rate and a relative cost value could, he explained, be understood through Einstein's equation $E = MC^2$. Einstein's equation provided "a single symbol—a unit symbol—of infinity itself, i.e., energy complete ('E')," which could be read at the basic rate of productivity.[173] Or, as described by his friend the Mexican sculpture Isamu Noguchi, "THE FORMULA THEREFORE PROVIDES A UNIT AND A RATE OF PERFECTION TO WHICH THE RELATIVE IMPERFECTION OR INEFFICIENCY OF ENERGY RELEASE IN RADIANT OR CONFINED DIRECTION OF ALL TEMPORAL SPACE PHENOMENA MAY BE COMPARED BY ACTUAL CALCULATION."[174] Taking the three variables through which energy accrued this relative value—"(1) Energy Conversion Ability, (2) Time, and (3) Precision"— Fuller translated Einstein's model of relativity into his own theory of *energy conversion* in application to price.[175]

By adopting the therblig standard proposed by the productivity experts Frank Gilbreth and Lillian Gilbreth, which reduced labor motions to the most efficient movements, Fuller called for a new energy yardstick—equivalent to the base yardstick rate provided by Roosevelt.[176] Taking the therblig standard to be like "Einstein's 'c^2,' i.e., TOP *efficient* speed" (or, in Roosevelt's case, a government backed cost), Fuller explained how these units form "a yardstick or basic rate in the terms of which the worker's TRUE dollar is interpolated: 'A dollar an hour' for basic energy conversion into work without recourse to initiative, which anyone can perform if scientifically instrumented and instructed."[177] Assuming this basic rate of conversion to be the benchmark rate, any "super-to-stint" activity that converted more energy would be additionally awarded.[178] Like the energy "yardstick," this claimed an absolute value for energy (a government price that accorded with Einstein's basic rate, or the Technocrat's "energy determinant"), while at the same time affording a relative price, correlated to the efficiency of its conversion.

Unsurprisingly, Fuller would have profited handsomely from this economic system. Having spent his career enhancing mobility forms, Fuller's prodigal ability as an "energy transformer" meant that he would have been a winner in his own economy of efficiency, rather than in the hard world of the dollar. Given his obvious personal bias, therefore, his principal criticism of Technocracy was that it stifled ingenuity; it understood energy only as a basic rate. It neglected the obvious fact that human will ("Passion, fashion, chance, change, intuition") remained in control of energy.[179] What the Technocrats excluded from their model was "SPECULATION and INITIATIVE in the acceleration of CHANGE." These, he argued, are "ALL-TIME FORCES, and are as essential in a scheme of realism as suffrage and the socialization of essentials and plenitudes."[180] Taking the notion of intrinsic value and transforming it into a model of "uncompromisable and untaintable dollarability," Fuller revealed the irony: in appealing to energy (as an objective model) to improve society, its value would always depend on the will, with its powers of initiative and speculation.[181] As a result, the "true rate maker of energy conversion" would be found in a reinterpretation of Einstein's famous equation, *Energy ∞ Man $\times$ Intellect*.[182] With the dollar being a "time-captured and saved unit of energy by man," it centered the human (as an energy convertor) in the production of value, while claiming "untainted" "teleological objectivity" for energy.[183] Fuller would, therefore, reduce energy to "dollarability," with energy being outside human intervention (and political speculation) while similarly being a residual product of man and intellect. Through this double maneuver, Fuller privileged energy over the dollar, objectivity over subjectivity, while collapsing the two together—energy being a product of "Man × Intellect." Fuller's scheme thus upheld the energy-money dialectic, while folding both into "dollarability."

Isabel Paterson and the Energy of the Free Market

To Fuller, despite its objectivity, energy would not stand outside the system of speculation: it centered the human in the process of valuation and as a result collapsed the dualism between energy and money. And yet, Fuller preserved this binary as he continued to privilege energy over money. Others used the logic of energy to naturalize money. The literary critic and staunch liberal Isabel Paterson put this argument forward most strongly. In her opinion, the further energy was equated with money, the more objective energy became. Energy, she recognized, was more natural when it was understood as money, rather than when it was positioned in opposition to it.

Paterson was a fierce opponent of the recent vogue in planning, calling for a libertarian movement based on laissez-faire capitalism. From her influential seat as columnist for the *New York Herald Tribune* "Books" section, Paterson criticized the fashionable intellectuals inspired by the fads of collectivism, communism, and state control.[184] As New Deal mania swept the nation, Paterson (affiliated with that now-unfashionable "classic American School of Thought") "suspected that she would be shot at dawn with equal enthusiasm by reactionaries, communists, fascists, internationalists, imperialists, technocrats, economists, and miscellaneous young radicals."[185] As the decade progressed, however, Paterson's mantra of ultimate freedom would be supported by a small band of followers who laid the grounds for the libertarian movement in the 1940s. Famously, she schooled Ayn Rand, twenty years her junior, on all aspects of her political philosophy.[186] Later, Rand bestowed the highest honor on Paterson's work, labeling her energetic ode *The God of the Machine* (1943) as "the basic document of capitalism."[187]

Despite her opposition to Technocracy, Paterson was attracted to energy's a priori appeal. Drawing on her unlikely friend Stuart Chase's "new philosophy of energy applied in space through time," Paterson adopted this mantra to undermine the New Deal vision it supported.[188] Whereas the energy-money Manichaeism supported Chase's New Deal vision of America, to Paterson money was energy, and this in itself was "a method of equating time and space."[189] In one of her famed *New York Herald Tribune* columns, Paterson explained why the dichotomy between energy and money had to be rethought. To her, the principal flaw of Technocracy was that it confused energy with "intrinsic goodness."[190]

Agreeing with the premise that the "enormous release of energy" would bring positive social benefits, Paterson refused to believe the supposition that magnitude was correlate to social good. John Ruskin had attached intrinsic value to moral directives such as aesthetics and social worth, but Paterson recognized that, in the absence of any axiology beyond energy itself, energy had no dictate to be intrinsically good. As such, there was a "terminological confusion" at the very heart of the Technocracy slogan "Wealth is a form of energy."[191] Even the phrase "wealth is the product of the expenditure of energy under intelligent direction," Paterson argued, was not without semantic upset: "A cocoanut might fall into the lap of a South Sea Islander who was too lazy to get out of the way; and yet that cocoanut would be wealth. And if the same primitive man were presented with a Carnegie library and the United States Steel Works, since he could not read and would have no use for even the best grade of steel rails, those things would not be wealth—to him."[192] Noting that there is no "self-existent economic system" beyond the human creation of value, Paterson assured that energy could only be valued according to what humans desired.[193] "It really depends," Paterson felt, "on what he wants and how much he wants it."[194] This itself would be a "natural" condition, for as the logic of laissez-faire capitalism postulated, when left alone the market (guided by Adam Smith's "invisible hand") functioned automatically, according to human desire. Value, in this framework, would be as natural as energy. Drawing on this association, Paterson affiliated herself with "energy" (backed by desire) rather than with energy for energy's sake. In order to do this, Paterson constructed an alternative economics of energy, centering the human (or money) in the naturalization of energy.[195] For Paterson, only through its conflation with money (rubber, or desire) did energy have any claim to objectivity.

Inspired by Henry Adams's *Education*, Paterson summed up a decade of thought on energy in *The God of the Machine*. In the intervening period, the Hitler-Stalin Pact had undermined the fashion for collectivism and government planning that characterized the early New Deal period. With the benefit of hindsight, Paterson reiterated her criticism of Technocracy and extended this to a critique of fascism and socialism, pointing out how magnitude on its own was not enough to coordinate a great civilization. The ancient world proved this: although Carthage had more energy, it was defeated by Rome. Quantity was irrelevant: what mattered (as Fuller had maintained) was the principle of direction. The recent influx of energy brought about by technology, Paterson argued in a portentous manner, would likely be disastrous: "a wrong hook-up more appalling in its effect by the given ratio, becoming apparent literally in a world explosion."[196] Energy had no value unless it was directed by spirit, in what Paterson coined the "long-circuit-of-energy." This was facilitated by money and allowed energy networks to expand in space and time. In opposition, idle money represented "a storage battery" and "a generalized mode of the conversion of energy when it is in motion, with a function of equating time and space."[197]

The difference between "Societies of Status" and "Societies of Contract," Paterson thus explained, depended on how energy was controlled, the former having a "short-circuit-of-energy" and the latter a "long-circuit-of-energy." The "long-circuit-of-energy" grew uninhibited, allowing energy to augment energy. In planned societies (such as New Deal America), due to artificial checks and controls, the "short-circuit-of-energy" gradually wound down, leading to a state of entropy. Paterson had already begun to develop this dichotomy in 1939 in her *Herald Tribune* column, where she praised the United States as a "dynamic" economy and characterized Europe as a "static" society.[198] In a "Society of Status," equivalent to fascist and socialist states, energy was inhibited by government controls and checks. In a "Society of Contract," the perfect liberal state, free will and intellect not only generated energy but also built structures that allowed the profligate flow of energy—the "long-circuit-of-energy"—to extend ad infinitum. This extension, she believed, depended on the freedom of a "First Creative Principle (God)" that undermined the second law of thermodynamics.[199] Even the despotic state, Paterson recognized, depended on this capitalist principle first, with "a few years of external force, the dictatorship of the proletariat or the élite, absolute government—and then no more effort, no need of intelligence, a machine running on—to a dead end."[200] In this way, Paterson claimed that "the theory of Marxist Communism is precisely that of the Perpetual Motion Machine, point by point, for it stipulates that the productive system created by free enterprise is a pre-requisite, to be taken over by the Communist machine."[201] As such, the laws of thermodynamics when taken alone, Paterson supposed, did not account for human intervention within an energy circuit, which "introduces a factor which does not answer to measure."[202] This intervention occurred in three ways: through the literal conversion of energy through the body, through intellectual capabilities, and through the "imponderable" acts that "*route* the energy he has tapped and brought under control."[203] In the perfect state, Paterson declared, energy required human intervention and desire to counter entropic dissipation: "left to itself, the water would never run uphill; it must flow down."[204]

Collapsing energy and money into the same semantic modality, Paterson reintroduced a binary between the structure of government and the energy that flowed through it. Energy as money was natural, but the structure that contained it was artificial, and if it encroached too far, it upset the natural movement of energy (money). Once again, therefore, Paterson had artificially positioned energy in opposition to the subjective checks of society. This was recognized by one reviewer in the *New York Times*, who pointed out that Paterson had fallen into the same intellectual fallacy as those who called for the energy theory of value.[205] In her attempt to "show that our republican institutions, as enshrined in the original Constitution and Bill of Rights, are so devised as to favor 'the long circuit of energy,' and are therefore responsible for the great material gains" of the past century, Paterson's "fundamental error" was her belief that government systems were a negation of energy—that they "are inert and have energies poured into them."[206] Governments were not, as Paterson so claimed, "end-appliances" to be fueled like a machine. "Energy is indwelling throughout nature."[207] As a result, man "is not differentiated from other things by being 'self-starting.'"[208] By positioning political forms as a negation of energy, Paterson had once again reinforced the subjective-objective divide. With energy and money indivisible, energy as money provided a more objective model than the government structure into which it was poured. Where the Technocrats used energy to naturalize the state against the subjective whims of the price system, Paterson correlated money with energy and, by doing so, presented the state as an artificial structure built on subjective whims.

Conclusion

In 1937, Harold Frazer provided a blueprint for an "energy certificate" in the official Technocracy journal, *Technocracy Inc*. Physically resembling the US dollar in all attributes except the information contained on the note, the energy certificate hardly differed from money.[209] With energy materialized in monetary form, the promise of energy came full circle. The desperate need for an economy backed by energy, so urgent during the difficult years of the Depression, came to a standstill, as it became evident that when abstracted as a thing energy was subject to the same problem as money: it remained a social relation. Despite promising an intrinsic, objective, natural standard, the moment energy was placed into circulation—found, for example, in the collection plate—it became a value like money, another standard to correlate incommensurable wants.

And yet, despite the obvious fallacy that energy was not a thing with an intrinsic value but a relation that gained its value through exchange, people were not put off its rhetorical promise but turned to it to naturalize the political ideology at the heart of the New Deal. Even once energy's objectivity became harder to defend as the decade progressed, Isabel Paterson, who understood energy to be money, would reinforce the subject-object divide as energy-as-money was placed in opposition to governmental checks, which controlled its flow. As Paterson dreamed of a liberal society free from the manipulative reins of the state, she, like the Technocrats, placed her faith in energy (the natural will of the people) rather than the inert mechanism of government, an empty form that had energy poured into it. Whether energy was opposed to money or operated as money was a moot point; it signaled an alternative to subjective whims, either of the market or, for Paterson, of the state. The "energy certificate" used against money (or as money) revealed itself as a convention, reinforcing the arbitrary boundary between "hardness" and "rubber," between objectivity and subjectivity. And yet, as Erginette had demonstrated so perfectly, these binaries broke down the moment energy performed the role of energy a little too well. The closer energy seemed to resemble the attributes that made it energy, the more it appeared to be a social relation, little different from money.

1939–1951

CHAPTER 5

Energizing the World

Energetic Ethics for an Atomic Age

"What the hell has Jeffersonian democracy got to do with it?" said Mr. Stoyte with mounting irritation. "Can't you believe in Jefferson and have your current wired in from the city?"
"That's exactly it," said Mr. Propter, "you almost certainly can't."
—Aldous Huxley, After Many a Summer Dies the Swan, 1939[1]

In August 1945, shortly after the United States dropped the atomic bomb over Hiroshima and Nagasaki, Norman Cousins, editor of the *Saturday Review*, penned an editorial titled "Modern Man Is Obsolete." In it he offered his apocalyptic dictum: "On August 6, 1945, a new age was born."[2] With Japan defeated and the Second World War coming to an end, America celebrated victory as trepidation grew over the new "cosmic force" unleashed. For Cousins "fear of the unknown, the fear of forces man can neither channel nor comprehend," marked the birth of the new era.[3] In the wake of mass destruction, Cousins believed that man's only possibility of survival was through an evolutionary overhaul—a planned intervention in the human—redirecting energy away from the repressive structures of individualism and nationalism toward greater levels of cooperation and "world conscience."[4] Although the recent United Nations Conference on International Organization, held in San Francisco in the spring of 1945, was a step in this direction, only the construction of a world consciousness, Cousins cautioned, would be strong enough to control the new cosmic energy released.[5] While scientists associated with the Manhattan Project called for the establishment of an international policing agency to control nuclear energy, prominent commentators, such as Cousins, understood that the solution to world government lay in returning to the problem of will—readdressing the shape and direction of human energies. Cousins's article hit a nerve: editors reprinted the copy, it was discussed extensively, and its phraseology entered popular discourse.[6]

Article after article stressed the necessity of a strong will in regulating this new cosmic force.[7]

In the coming weeks, journalists struggled to explain atomic energy to a bemused public, describing how the atom was split, muddling through Einstein's equation $e = mc^2$, and assuring the reader that atomic energy was still part of the cosmic continuum. Cousins's editorial, however, shows how, despite new scientific paradigms, earlier conceptualizations of energy remained in public discourse. Attention had once again turned to the channels of energy, to a consideration of its flow through society—to its redirection by the will. Indeed, an editorial published in the Saturday Review two weeks later, titled "Ethics for an Atomic Age," revealed how Cousins had reverted to a familiar conceptualization of energy to make sense of this new cosmic force.[8] Under the title "Ethics for an Atomic Age" sat an extract from Thomas Henry Huxley's "Evolution and Ethics," first delivered as a Romanes Lecture in 1893. In his address, Huxley described how human energy, encapsulated by the "ethical process," was working against the "cosmic process at every step."[9] Foregrounding the importance of a strong will over nature, the republication of Huxley's lecture reminded a nervous postwar audience that "fragile reed as he may be, there lies within man a fund of energy operating intelligently" in the universe, able to refashion the "cosmic process" to its own end.[10] At the start of Cousins's "new age," the Saturday Review turned to "Darwin's bulldog" to clothe an atomic ethics in a nineteenth-century evolutionary moralism. Huxley's vision of nature as a garden tended by man provided the Saturday Review with a founding metaphor for the post-1945 era, its manifestations of cosmic energy "alternating between phases of potentiality and phases of explication," waiting patiently for a strong will to harness it into form.[11]

Not only did Cousins's editorial owe a considerable amount to Thomas Huxley's understanding of the ethical process, but it drew extensively on (and directly quoted) the work of his grandson, Julian Huxley.[12] Like his grandfather, Julian Huxley had spent his career as a scientist, science popularizer, and social commentator, exploring the relationship between ethics and the evolutionary process. A founding member of the planning group Political and Economic Planning (PEP), a staunch advocate of New Deal planning, and the first director general of the United Nations Educational, Scientific and Cultural Organization (UNESCO), Julian Huxley redefined ethics in practical terms as a social and biological tool to achieve a more rational society. He updated his grandfather's talk in his own Romanes Lecture, "Evolutionary Ethics," given fifty years later in 1943, with the question of postwar reconstruction on everyone's mind. For Julian Huxley, the "proto-ethical mechanism" could be understood according to the same architectonics of the psyche: "it charges all that passes through its mill with the special emotive qualities of rightness or wrongness."[13] Drawing on his long engagement with Freudian psychoanalysis, he redefined ethics as a structural phenomenon that extended like a Freudian superego ("a supporting psychological frame-work") across society.[14] In a continual process of evolution, this plastic armature, Huxley believed, could harness and reengineer society's energy toward the socially productive ends needed for evolutionary progress. In his talk, he even turned to the example of William James's "moral equivalent" as a prototype of how to affect this wholesale redirection of energy.[15] For the purpose of postwar reconstruction, ethics, Huxley maintained, would provide the necessary armature to channel energy away from laissez-faire capitalism and the narrow confines of nationalism toward more equitably distributed wealth and greater international unity, as promised by the ideal of the world state.[16]

Cousins's debt to Huxley shows how, at the dawn of the atomic age, the same energy paradigms continued to dominate the public imaginary. Energy was once again envisioned as a fluid substance harnessed through society by the will, similarly to how water moved through a watercourse. Scholars such as Paul Boyer, Alan Nadel, and Margot Henriksen have documented the emergence of an "atomic consciousness" in the post-1945 period.[17] This work has captured the myriad ways in which American culture and society were transformed by the threat of the atomic bomb. This "atomic consciousness," however, did not extend to the way that energy itself was conceptualized. Even though the atomic age came with new scientific explanations of energy, it did not radically transform the way that energy was discussed in public discourse. Rather, as this chapter will demonstrate, the same energy paradigms that grew out of the Progressive Era continued to structure how energy was conceptualized well into the atomic age.

These frameworks, however, evolved to reflect the changing political landscape. In 1939, the Hitler-Stalin Pact undermined the enthusiasm for rational planning that had dominated the early New Deal period, as the Soviet Union aligned itself with fascism. In opposition to regimes in Italy, Germany, and the Soviet Union, democracy became a rallying cry against the threat of totalitarianism.[18] Faced with this threat, intellectuals fixed an image of American democracy as the opposite of totalitarianism.[19] Progressives who before the war had celebrated Marxism along with New Deal reforms suffered a crisis of faith as these became increasingly associated with totalitarianism. The former Trotskyist Dwight MacDonald plotted the shifting coordinates of the Left through his journal *politics*, as it came to privilege the individual above the overbearing forms of mass society.[20] Alongside the rise of an increasingly individualistic politics there appeared a growing neurosis, fueled by the work of the Frankfurt School, that totalitarianism could be homegrown, abetted by mass consumerism. In light of these anxieties, intellectuals such as Lewis Mumford, once the voice of planning, also began to echo the ideals that were being promoted in mainstream politics, turning his attention toward rebuilding man.[21] Rather than looking toward the collective energy of the watercourse, intellectuals turned inward, adopting ethics as a tool to concentrate society's energy and strengthen the boundaries of the self.[22]

By tracing a discourse centered around what I shall term *energetic ethics*, this chapter will capture how earlier energy paradigms matured in the late 1940s to reflect the changing political landscape of the atomic era. The chapter will start by tracing how energy became central to a wider cultural debate about rebuilding the "new man." This was best represented by Lewis Mumford's evolving social thought, as well as by the growing status of the individual within the cultural zeitgeist of the period, exemplified through the painting of abstract expressionist Jackson Pollock and the writing of the poet Charles Olson. At the heart of this was a call for the construction of stronger ideational structures (such as ethics, religion, and mysticism) to harness and reshape society's energy. The call to better organize society's energy, in turn, contributed to a growing paranoia about new forms of totalitarian control emerging within American society and mass culture. To those who escaped fascism in Europe, such as the members of the Frankfurt School, any discussion of managing energy appeared dangerously close to the restrictions placed on energy within a totalitarian state. The debate, therefore, extended to considerations about whether energy should be liberated from all social constraints or carefully managed, albeit framed through discussions about freedom and social control.

Just as energy provided a model to look inward, the second half of this chapter will examine how it became a way to reimagine the political order of the post-1945 world. Julian Huxley, for example, called for a new energetic ethics to refashion the self and lay the intellectual foundations for one of the leading postwar international organizations, UNESCO, as its first director general. This chapter, therefore, will follow how energetic ethics fed into the language of individualism, while at the same time becoming a way to reimagine a unified world state.

The connection between energy and moral value once again sat at the heart of these debates. Even before atomic energy emerged as an apocalyptic specter hanging over the postwar era, fears over the irrationality of man—exposed by the rise of Nazism and the horrors of war—continued to fracture any moral claim that energy had gained over the past three decades. This was best exemplified by

Wilhelm Reich, who had once been the most vocal advocate for the moral priorism of energy. By 1951 Reich began to identify some negative life force in operation, which he called "deadly orgone," or "Dor" for short. He had discovered this negative life force as he conducted experiments with radium to discover whether orgone could neutralize the effects of atomic radiation. As these experiments only heightened radiation levels, Reich identified a deadly energy in the atmosphere, which he believed caused nausea, chronic fatigue, and intense cramping and even blackened the rocks near his house.[23] Following this, the dark atmospheric clouds that Reich attempted to dissipate with his invention of the "cloud-buster" became symbolic of how energy, which had once appeared a stable yardstick (a measure not only of American civilization but also of man, culture, and the economy), stood as a mushroom cloud over the postwar era in desperate need of form. With energy's moral privilege compromised, many agreed with Julian Huxley that the optimum "yardstick" for democratic society was not energy but an "internal yardstick" that measured the individual's capacity to harness energy within the political structure demanded of the new world. This would be a yardstick based on form rather than force.

Lewis Mumford's Blueprint for a "New Man"

On October 17, 1944, Lewis Mumford and his wife, Sophie, received the telegraph that thousands of other families in America awaited with dread. Their son, Geddes Mumford, had been missing in action and would not be coming home.[24] A devastated Mumford found solace in writing about his son and penned his most unpopular book, *Green Memories: The Story of Geddes Mumford* (1947). In this book, Mumford reflected on how his son had grown into the man he had become and what institutional forms (including parenthood) had shaped him. The only consolation he could find in his loss was the possibility of reinventing man and changing the foundation of the world his son was born into.

Almost seventeen years after *Technics and Civilization* was published in 1934, Mumford wrote the last book in his Renewal of Life series, *The Conduct of Life* (1951), laying out his blueprint for a new man. Mumford had been developing this thesis since the beginning of the war, but *The Conduct of Life* stood as its apotheosis. The book took as its starting point the dismal position modern man found himself in. Never before in human history had mankind had such an abundance of "vitality and energy" on tap.[25] Yet, he asked, "what use are cosmic energies, if they are handled by disoriented and demoralized men?"[26] For too long, Mumford explained, Western civilization had been under the spell of power, forgetting that "uncontrolled power in any of its manifestations, as heat, as light, as physical force, as political compulsion, is inimical to life."[27] Instead, Mumford argued that the value of life depended on the extent that it could "regulate power" and mold energy into form.[28] Lacking this capacity to form, civilization had fallen into barbarism. As it moved ever closer toward "disintegration," it had produced "mass man": "a creature governed mainly by his conditioned reflexes—the ideal type desired, if never quite achieved, by the advertising agency and the sales organizations of modern business, or by the propaganda office and the planning bureaus of totalitarian and quasi-totalitarian governments."[29] Mumford's message was clear: "All the resources our society now possesses, all its present energies and vitalities, all its funded values and ideas, must be concentrated on the upbuilding and regenerative functions, in both the personality and the community."[30] Only through an axial shift in values, Mumford believed, could energy be concentrated toward those forms needed to protect modern man from the encroaching threat of totalitarianism.

Though it culminated in *The Conduct of Life*, Mumford's call for an integrated value system had been developed during the early years of the war, as he emerged as one of the most vocal liberals to advocate American intervention in Europe. In a confrontation that rehearsed the Bourne-Dewey schism of the previous war, Mumford attacked the political apathy of his fellow intellectuals as exemplified by their support of the current government policy of isolation. In 1940, he resigned from his post as commissioning editor of the *New Republic*, accusing the other senior editors, Bruce Bliven, George Soule, and Malcolm Cowley, of moral cowardice for refusing to confront the threat of fascism head-on.[31] As he departed, he outlined his political protest in an article titled "The Corruption of Liberalism," which offered a neat summation of all he felt had gone wrong with the liberal creed. In the article, Mumford identified two types of liberalism that had split the Left: "ideal liberalism" and "pragmatic liberalism." Setting himself in opposition to pragmatic liberalism, he accused this group of giving up on "esthetics, ethics and religion," of abetting a form of "emotional anesthesia" that had eradicated all moral coordinates from the world.[32] Placing faith in the innate goodness of the human subject to counter the forces of barbarism, the "pragmatic liberals," Mumford charged, had once again privileged energy over form, the "machine" over "moral values."[33] To oppose this, Mumford called for a revival of ideal liberalism to bring about "a recrystallization of the positive values of life, and an understanding of the basic issues of good and evil, of power and form, of force and grace, in the actual world."[34] Echoing Bourne, Mumford argued for a wholesale reinvestment in moral value as the primary instrument to regulate human energy.

Having begun intensive research into early Christianity in the winter of 1939/1940, Mumford placed the onus on religion to bring about this spiritual revival.[35] He advocated a nontheistic religion in which social, moral, and religious values would be united. In *The Conduct of Life* Mumford explained how this religion would operate primarily as an organizational principle: a means to structure the expenditure of energy.[36] In his previous book, *The Condition of Man* (1944), Mumford had provided a historical chronicle of how symbolic forms had organized the direction of society's energy. Here he charted the historical formation of the subject out of the community and the idolum in which they were "enlarged, energized—and completed."[37] By tracing the evolution of the idolum, Mumford underlined the role of the symbolic plane in structuring internal and external energy patterns.

Baroque art, to Mumford, represented the clash of two great idola, or energy patterns: Protestant Catholicism and medieval Christianity.[38] According to Mumford, Protestantism, following Max Weber's formalization, provided the authoritarian channels to control man's physical and moral energies toward the productive ends demanded by capitalism. The lust for life produced by the medieval period, on the other hand, resulted in such "sheer animal vitality" being "translated into such a wealth of esthetic and practical forms" as had never been seen before.[39] Baroque art, for Mumford, stood as the union of these two competing social orders: between the sixteenth and nineteenth centuries, just as the "life-denying forces" of capitalism grew, so too did "an uprush of the libido," a "floodtide of vitality that spread into every cove and channel of the spirit."[40]

In *The Conduct of Life*, Mumford called for modern society to organize its energy, concentrating it toward the building up of life. Echoing Brooks Adams's "Law," he pointed out how there had been moments in history where there had been great leaps forward in civilization, examples of a "sudden concentration of energies, a more favorable constellation of social opportunities, an almost worldwide upsurge of prophetic anticipation."[41] The discovery of the bomb had made this focus imperative, and as a result, Mumford placed his faith in form, in the structure of ethics and religion, to control and mold the new energy released.

The Hero's Energy

While Mumford placed the onus on ethics and religion to channel society's energy, others placed the responsibility on the individual. The heroic individual took on increasing prominence within the zeitgeist of postwar America as the collective politics of the New Deal waned. Within this framework it would be the individual hero who would channel and redirect society's energy, rather than collective forms of organization such as the watercourse.

For his image of civilization taking great strides forward through the concentration of energy, Mumford was indebted to the British historian Arnold Toynbee.[42] In 1947, an abridged version of Toynbee's history had been published in the United States to wild acclaim, making an unlikely celebrity of a British academic steeped in Edwardian culture. At this time, six of the twelve volumes had been published, three in 1934 and a subsequent three in 1939. However, where these earlier volumes were directed toward a scholarly audience, the abridged version completed by the English schoolmaster D. C. Somervell became an immediate bestseller, transforming Toynbee into a cause célèbre. Whether many readers sat down to digest Toynbee's convoluted study in its entirety was not the point; the book appealed to its audience, and people bought it. With Toynbee's photograph appearing on the front page of *Time* magazine in 1947, it was clear that *A Study of History* suited the mood of postwar America, just as Spengler's *The Decline of the West* offered a mirror to society in 1929.[43]

In 1929 Spengler had fixed the will into the organic life cycle, eliminating agency from the cosmic process. He popularized an essential primitive energy running underneath an artificial civilization. Yet as the United States took its place on the world stage, Toynbee stressed the role of the will in building society, in taking control of its energy. In September 1947 an article by Toynbee published in the *New York Times* reiterated the importance of the human spirit in ensuring civilization's survival.[44] Influenced by Ellsworth Huntington's pulsatory hypothesis, he pointed out that, even though civilization was tied to organic cycles, through "creative action" civilization could escape the deterministic fate of "social suicide," whose potential had been heightened by the bomb.[45] "We are endowed with this freedom of choice," Toynbee told his readers, and we "cannot shuffle off our responsibility upon the shoulders of God or nature. We must shoulder it ourselves."[46]

Despite this optimism, in *A Study of History* Toynbee predicted limited odds for Western civilization. Out of twenty-one civilizations, he argued, only Western civilization remained. To plot this progress, Toynbee collapsed civilizations on top of each other, depicting them as climbers on a mountaintop, with "primitive" societies "lying torpid upon a ledge on a mountain-side" and higher civilizations "likened to companions of these 'Sleepers of Ephesus' who have just risen to their feet and have started to climb on up the face of the cliff."[47] Civilizations could ascend the mountain through self-determination, as they managed to surmount

Illustration of a broken Grecian bust representing Arnold Toynbee's theory about the decline of civilization. Hoyningen-Huene, *"The Dead Hellenic World": Roman Statue in Corinth*. In "Is Our Civilization Doomed? Explaining the Death of Other Cultures, Mr. Toynbee Holds Out Hope for Our Own," *New York Times*, April 13, 1947, 203.

external obstacles by mustering an innate drive. This came about through the reciprocal dynamics of "challenge" and "response": growing "an *élan*" that led in the "Macrocosm" to a "progressive mastery over the external environment" and in the "Microcosm" to "a progressive self-determination or self-articulation."[48] The act of self-determinism was brought about by a process Toynbee termed "etherialization," a state that led to heightened efficiency, by liberating "forces that have been imprisoned in a more material medium," set "free to work in a more etherial medium with a greater potency."[49] Toynbee's process of etherialization thus evoked a channeling that required "a consequent transfer of energy, or shift of emphasis, from some lower sphere of being or of action to a higher one."[50] The progress of civilization, Toynbee reiterated, depended on the potential to sublimate energy away from practical matters of survival toward an elevated spiritual plane. This was not, however, a collective endeavor, as it had been for other prophets of civilization. To reach this higher state, energy had to be harnessed by an individual hero, a "creative personality," who led civilization further up the mountain. Etherialization, in this sense, was a two-way process. The individual channeled energy toward a higher spiritual plane and thereby overcame the inertia of his fellow men. Through a reciprocal process of *"mimesis,"* society's energy would follow this individual's direction toward a higher medium.[51]

In this model, Toynbee positioned the "creative individual" at the heart of an energy system, out of which the community gained its form. This centering spoke to the growing prominence of the individual as the collective politics of the New Deal waned. In a climate where the individual assumed greater visibility, others turned to Toynbee's study to identify examples of the willing individual, a figure able to harness and mold society's energy. By 1949, for instance, the comparative mythologist Joseph Campbell mapped the model of the hero through the ancient myths of different cultures. Drawing on Carl Jung's concept of the archetype, Campbell provided in *The Hero with a Thousand Faces* a structural model of the hero's monomyth within a three-stage process of *"separation— initiation—return,"* leading to the transformation of society.[52] This three-stage process hinged on the hero's "return and reintegration with society," for this was the moment when the hero's transformation enabled the "continuous circulation of spiritual energy into the world."[53] Throughout the book Campbell described the hero as an energetic figure "filled with a double charge of the creative energy," someone who retrieves a "miraculous energy-substance" from the Gods in order to recirculate it back into the community.[54] The rewards of the hero's quest, "the ultimate boon" and the "female lover," were conduits of a nourishing "life-energy" that had to be returned to the community.[55] For the "hero" to be successful, he had to complete his task before the "unlocking and release again of the flow of life into the body of the world."[56] The hero thus had the responsibility of reinvigorating a community's energy: through his trial he circulated energy back through the social body. Reversing the dynamic whereby it was "society that is to guide and save the creative hero," in Campbell's narrative the hero undergoes a spiritual transformation and thereby funnels energy back into the community, initiating its spiritual conversion.[57]

In 1951, a short essay by the poet Charles Olson titled "The Gate and the Center," published in the avant-garde journal *Origin*, revealed how the archetype of the energy-wielding hero would shape the imagination of postwar American modernism. During the war, Olson had been a New Dealer working in the Office of War Information, generating propaganda for the war effort.[58] After the war, Olson frequently visited Ezra Pound, who had been charged with treason and was incarcerated in St. Elizabeths Psychiatric Hospital in Washington, DC. Expanding on the Pound/Fenollosa principle of kinetic verse, Olson penned his manifesto "Projective Verse" in 1950, formulating a new "OPEN FIELD" form of poetry that depended on the principle that the poem "must, at all points, be a high-energy construct and, at all points, an energy discharge."[59] The following year, Olson expanded his model of "OPEN FIELD poetry" into a cultural statement about the role of the artist in society. Echoing the concern over the lack of form in contemporary society, Olson described the importance of "CENTER," "SHAPE," and "FORCE" for the preservation of civilization, using the earliest known civilization, ancient Sumer, as an example of social cohesion.[60] Calling once again for greater concentration, Olson asked the reader to imagine what man could be "once the turn of the flow of his energies that I speak of as the WILL TO COHERE is admitted, and its energy taken up."[61] After all, Olson went on, as "energy is larger than man," if the hero "taps it as it is in himself, his uses of himself are EXTENSIBLE in human directions & degree not recently granted."[62] He continued, "The EXCEPTIONAL man, the 'hero,' loses his description as 'genius'—his 'birth' is mere instrumentation for application to the energy he did not create—and becomes, instead, IMAGE of possibilities implicit in the energy, given the METHODOLOGY of its use by men from the man who is capable precisely of this, and only this kind of intent & attention."[63] The archetype of the hero thus centered the individual, in Olson's words, as the "gate" through which nature's energy was shaped and brought into form. Energy would be channeled through him, as the hero and the universe were united in a creative act.

Nowhere would this sentiment become more evident than in abstract expressionism, which emerged as America's dominant aesthetic style in the late forties. Indeed, one of its principal critics, the writer Harold Rosenberg, had characterized it in exactly these terms, as he designated the new aesthetic style "action painting."[64] For Rosenberg, what was unique about the new American style was that for the first time the content of these works reflected "the way the artist organizes his emotional and intellectual energy as if he were in a living situation."[65] In the hands of this distinctly "American School," the painting had become a space to act rather than to "reproduce" or "re-design" an object.[66] It had become an "event," a record of the artist's own psychic conversion and interaction with the world. Within this new school of painting, Rosenberg noticed, the spatial had given way to the temporal, leading to a new "vocabulary of action": "inception, duration, direction—psychic state, concentration and relaxation of the will, passivity, alert waiting."[67] The painting had become, he described, a three-dimensional space in which the artist would organize energy. As artists came to identify their work along these lines—with Robert Motherwell describing it as "an energetic field of force" and Jackson Pollock describing it as "energy and motion made visible"—the artist was privileged as a nodal point through which energy was concentrated.[68] In 1949, as *Life* magazine declared Pollock (albeit in tongue-in-cheek fashion) America's greatest living painter, it appeared that there could be no higher praise bestowed on a work of art than it being an "outpouring of Herculean energy."[69]

Totalitarian Energies

The call to redirect society's energy, through both new ethical structures and the hero's actions, however, was attacked by critics, who viewed this as encroaching dangerously close to social control. After all, attempts to manage society's energy had found a prominent place within the rhetoric of totalitarian regimes, such as the Soviet Union. Energy, therefore, fed into wider intellectual debates about the spread of totalitarianism around the world and the danger of it surfacing within the United States.

Mumford's appeal to reorganize society's energy had been set against the "pragmatic liberals," whom he accused of facilitating fascism by failing to concentrate energy or mold it into form. In this plea, nevertheless, Mumford was also accused of enabling fascism, with one critic even charging him of putting forward "the most flagrant statement of the 'liberal call' for fascism so far."[70] From 1938 onward, different factions of the Left took aim at Mumford's quasi-spiritualistic prophecies, as he was increasingly satirized as the figurehead of a defunct liberalism. One article in the socialist mouthpiece the *New Masses* went so far as to claim that he was "Nazifying America," comparing passages in *Faith for Living* with Hitler's *Mein Kampf*.[71] The writer criticized Mumford for neglecting the forces of the class system and glorifying a social system built on "a new joy in fecundity."[72] How dare Mumford, the writer asked, put the blame on the "wicked 'fashion'" of the "dress-makers and typists" for pursuing a comfortable living, rather than husbanding "their energies for nurture and for passionate play."[73] Not only was Mumford unfairly moralizing the worker's energy as people struggled to make a living, but the writer also accused him of fighting fascism with fascism. In his attempt to reorganize the energies of the democratic individual, Mumford was accused of using the same totalitarian mechanisms of control as fascism. He had become a high priest lecturing the public on how to use their energy.

This argument erupted in 1943, with both sides accusing the other of abetting fascism. The counterattack came from the "pragmatic liberals" surrounding the *Partisan Review*, who diagnosed the idealistic, metaphysical, and pseudoreligious thought proposed by Mumford and other close affiliates as part of a "New Failure of Nerve."[74] Mumford would be the target of this attack, along with his friend Reinhold Niebuhr, Arnold Toynbee, the poet Archibald MacLeish, and Julian Huxley's younger brother Aldous. Sidney Hook, a devoted pragmatist and former Marxist (nicknamed "Dewey's Bulldog"), led the debate, to be later joined by John Dewey and Ernest Nagel. "Into the breach," Hook charged in an opening salvo, "has stepped the motley array of religionists filled with the *élan* of salvation and burdened with the theological baggage of centuries."[75] In the wake of economic crisis, world war, and totalitarianism, the "interpreters of divine purpose" "have now become concerned with social healing, with the institutions of society and with the bodies of men, as necessarily involved in the healing of individual souls."[76] Rejecting positivist science and rationalism, these thinkers, Hook believed, celebrated a "rise of asceticism, of mysticism" and "pessimism," the "frenzied search for a center of value that transcends human interests," and the rise of "theological and metaphysical dogmas."[77] In their search for a center of value, these thinkers had placed new controls over society's energy.

The increasingly personal attacks directed against Mumford during this period demonstrated that although the call to harness energy fit the progressive mood of the New Deal, by the end of the decade the same appeal appeared suspect. This was not helped by the shared language that had once been used to describe the Soviet Union, along Stuart Chase's lines, as an enlightened model of how a society's energies could undergo a complete overhaul.[78] Not only had technocratic progressives like Chase fawned over the economic efficiency of the Soviet Union's planned economy, but during the peak of its popularity in the early years of the Depression the Soviet Union was celebrated as an experiment in how all forms of energy—social, sexual, and cultural—could be organized on a new material basis.[79] John Dewey, for instance, who had made the pilgrimage to Soviet Russia in the summer of 1928 (as head of a group of twenty-five prominent American educators), would describe his experience of Moscow as a "restless movement, to the point of tension," that was the result of "a creative energy that is concerned only with the future."[80] This energy was directed to the point of an "*élan*," charged with "the ardor of creating a new world."[81] He described Moscow as "the heart of the energies that go pulsing throughout Russia," praising the Soviet Union as a new society founded on "a sense of the planned constructive endeavor which the new régime is giving this liberated energy."[82]

Dewey was not alone in celebrating the Soviet Union as a great emancipator and controller of energy. During this period, the press carried many laudatory accounts written by American tourists who had traveled to Russia and represented it in these terms.[83] In 1930, for example, the journalist Louis Fischer, then living in Moscow, described for the *Nation* his impressions of the Soviet Union as being hit by a "newly released wave or wall of energy and enthusiasm."[84] The communists, he recognized, had "become machines of permanent motion" working with a fiercely goal-oriented energy: they were harnessing it toward "the destruction of capitalism in the Soviet Union."[85] In comparison to a depression-ridden America, "everything moves here. Life, the air, people are dynamic. When I watch these recently unsealed reservoirs of energy I am sometimes carried away and think that nothing is impossible in the Soviet Union."[86] Thus, while, during the early years of the five-year plan, leading intellectuals such as Dewey could celebrate the "Russian experiment" as a great harnesser of energy, enthusiasm had waned by the late thirties.[87] As more information was fed to the American public about severe food shortages, high levels of political repression, terrible living conditions, and political show trials, all topped off by the political fallout brought about by the Hitler-Stalin Pact, by the end of the decade any attempt to overhaul society's energy would appear to be another tool of social control.

This growing sentiment was best articulated by free market liberals such as Isabel Paterson, who celebrated an unshackled energy in *The God of the Machine* (1943), freed from the constraints of the state. Her mentee Ayn Rand went even further in *The Fountainhead* (1943), aligning her philosophy of egotism with a self-generating energy that could escape any attempt at control, including the physical limitations dictated by the laws of thermodynamics. Rand, a Russian émigré whose family had suffered in the Soviet Union, harbored a powerful hatred of communism and had been horrified by the infatuation with collectivism that saturated the intellectual community of New York. She subsequently developed her own philosophy of "individualism" as a counterpoint to the political climate of the 1930s. *The Fountainhead* (1943) would become a monument of her individualistic philosophy, and she sold a hundred thousand copies in 1945 alone, topping the *New York Times* best-seller list. In the novel, Lewis Mumford was satirized as the demon critic and collectivist Ellsworth M. Toohey, whose tentacles reach through the novel as he gains power while preaching selflessness.[88] Throughout *The Fountainhead*, Toohey does everything in his power to dam energy, to stop it from flourishing from Rand's individualist hero, the architect Howard Roark. In contrast to and defiance of this control, Roark is depicted as a dynamo that pulsates energy "to all those around him" "through the walls of his office" to the city beyond.[89] Not only does Roark radiate energy through the city and the novel, but his defense of individualism given in the climactic courtroom scene also defined egotism in energetic terms: "it is the whole secret of their power," Roark explained, "that it was self-sufficient, self-motivated, self-generated. A first cause, a fount of energy, a life force, a Prime Mover."[90] While Roark served as a powerful dynamo radiating energy to those in his vicinity, "second-handers," such as Peter Keating, were characterized as "motors" who required constant "refueling" from others.[91] These "second-handers" depended on the individualist's energy, while Toohey governs the entire social structure, in his words, as "a single heart, pumped by hand"—his hand.[92]

Not only did *The Fountainhead* put forward a strong case for a liberated rather than controlled energy, but as Toohey, architectural critic on the daily *Banner*, fed energy back into the system—manipulating public opinion, aesthetic style, and social values—the novel also articulated wider anxieties over the control of society's energy by mass culture. This growing concern not only was explicit in Rand's social philosophy but also was reflected in a widespread concern as to whether thought structures could be influenced as well, resulting in a proliferation of studies on the subject.[93] At the heart of this national self-analysis sat the question of durability: was the American vulnerable to the same authoritarian thought structures that had swept through Europe?[94] Members of the Frankfurt School had taken up this subject, having relocated to Columbia University after Hitler's rise to power. Looking on with horrified fascination at the ubiquity of popular culture in America—its films, advertising, sports, and radio—members of the school (most notably Max Horkheimer and Theodor Adorno) turned their attention to the workings of the "culture industry" to expose the hidden authorities at work.[95]

The awareness that mass culture was manipulating thought patterns led to a discussion as to whether energy was bound to institutional regulation—controlled by the internalized structures of society. The German émigré and psychoanalyst Erich Fromm took up this subject, as he turned to Reich to determine whether energy patterns were historically conditioned by the internal pressures of society. To Fromm, who was a member of the Frankfurt School until he split with the group in 1936, the freedom promised in a democratic society was illusory.[96] Instead, capitalism had its own system of authoritarian control, as it transformed "mass-man" into "automatons who live under the illusion of being self-willing individuals."[97] In Fromm's view, the institutional structures embedded within corporate America shaped energy into the necessary channels needed for the smooth functioning of society. Because of this, Fromm described how the role of social character was to determine the "form in which human energy is shaped by the dynamic adaptation of human needs to the particular mode of existence of a given society."[98]

Having grown up an orthodox Jew, Fromm had experienced the impact that thought structures, like religion, had on character formation.[99] He had even taken up this subject in one of the Frankfurt School's first empirical studies, as he profiled the psychological makeup of the German working class. Although this (along with other practical studies) was prematurely aborted, Fromm's 1941 book *Escape from Freedom* (published outside North America as *The Fear of Freedom*) reached a wide audience, going through five print runs during the war. Challenging Freud's biological grounding of the libido theory, Fromm maintained in *Escape from Freedom* that patterns of energy were molded through history by society: the medieval serf channeled energy through a different pattern series from the Reformation man and the modern subject. Given the socially determined nature of this process, Fromm popularized the notion that authority was not imposed from outside; rather, *"the social character internalizes external necessities and thus harnesses human energy for the task of a given economic and social system."*[100] This is why, Fromm argued, Protestantism had been such a productive structure for capitalism, for "the inner compulsion to work" demanded by Calvin's dictum to submit entirely to God "was more effective in harnessing all energies to work than any outer compulsion can ever be."[101] Protestantism, in Fromm's diagnosis, carved out the "specific forms into which human energies was shaped" and subsequently became "one of the productive forces within the social process."[102] Aligning "religious, philosophical or political" thought as part of the same social process, Fromm pointed to the functional part played by the ideational plane in controlling energy.[103] Because the modern subject craved authority, the only way to escape fixed energy patterns was through "spontaneity," through an enhanced freedom of the subject.[104] Freedom could be brought about when the individual was able to locate a greater relatedness to the world. This, Fromm believed, would be engendered through the careful planning and control of the material basis of human existence.

Liberated Energy and the Paradox of Control

Energy, in Fromm's view, was socialized, conditioned as it was by shared institutions and history. Yet while Fromm stressed the institutional nature of energy, other political radicals, seeing rational organization of the socioeconomic structure as the road to greater freedom, turned to Wilhelm Reich's theory of "orgone" to maintain that energy had a political reality *prior* to form and thus could become a template to escape institutional forms altogether.

At the heart of this movement was the psychologist, libertarian anarchist, and enfant terrible of the Greenwich Village scene Paul Goodman, who absorbed Reich's model of orgone energy to advocate the destruction of repressive social institutions. This was in direct opposition to what he perceived as Fromm's solution of social engineering, which could be rejected along the same political lines as the "New Deal, the Beveridge Plan, Stalinism, etc."[105] Drawing on the Comtian term *"sociolatry,"* a "religion of society," Goodman described the alienation wrought by contemporary society, where the "natural energies are absorbed, sublimated, and verbally gratified in our corporative industrial states."[106] Energy, he alleged, was trapped within an institutional nexus controlled by mass culture. Submerged underneath this "sociolatry," Goodman held, was a universal energy, a power "immeasurably stronger than those alien institutions," which could rebel "against the superficial distractions of the ego" to liberate it from the inanity of modern institutions.[107] As this energy erupted from the essential nature of man, "orgastically potent" people, Goodman hoped, would shed the institutional forms of industrial society and as a result create novel solutions to the present situation based on their instinctual drives.[108]

Although Goodman's "gonad theory of revolution" later inspired a select group of dissident intellectuals (led by Dwight Macdonald, Norman Mailer, William Burroughs, and Allen Ginsberg) to celebrate sexual energy, promoting naked cocktail parties, orgies, and promiscuous behavior as part of a social revolution, others remained cynical about the political potential of privileging this energy.[109] Furthermore, in situating freedom as an essential energy flowing beneath society, Goodman was criticized for delimiting choice by reducing society to an absolute value—to an essential energy. One of the sharpest critiques of the "gonad theory of revolution" was launched in the journal *politics* by the sociologist C. Wright Mills and Patricia J. Salter. They charged Goodman with confusing "orgastic potency" with morality and masking "ethical choices by a metaphysics of biology."[110] Goodman, they argued, had used energy as a common measure that assumed commensurability between disparate fields—between politics, economics, and sex. There was no direct relationship, however, Mills and Salter pointed out, between these diverse sets of experience. Due to the complexity of social life, not only would orgone reduce a broad range of sexual practice to one reductive drive, but it collapsed the multiplicity of economic and political experience into an equally standardized form.[111] By privileging energy as prior to politics, Mills and Salter—adopting Fromm's argument—accused Goodman of having neglected the fact that "impulses are given *content* only by the participation of men in given institutions."[112] Freedom, they pointed out, did not come from reducing value to one biological necessity—to a primal energy—but in the opening up of new values and possibilities within society.[113] Freedom, instead, depended on creating new channels and social institutions into which energy could flow.[114]

How energy was positioned in relation to social institutions thus took on added importance as fears grew over totalitarianism. On the one hand, as Mumford argued, for the democratic subject to fight fascism, energy needed form; it had to be channeled toward the values required to rebuild modern man. From another perspective, any attempt to regulate, pattern, or harness energy could be equated with authoritarian methods of control. At the same time, to anarchists such as Goodman, it was only by reengaging with an essential energy, with an energy that preceded form, that the human could be liberated from the oppression of corporate society. However, critics of this position recognized that by submitting to an absolute energy freedom would be curtailed altogether.

This ambiguity, which meant that energy could aid in the construction of the autonomous subject and at the same time eliminate autonomy, pushed the energy paradox to its logical conclusion. Only by husbanding energy could one escape institutional controls, but to find an escape from these authoritarian forms, new institutions—symbolic, religious, and ethical—had to be erected to direct energy toward the social values required to meet this end. Freedom, from this angle, could only be achieved by further disciplining energy—by placing further restrictions on its expenditure. One critic traced the irony of this position within the growing trend for mysticism that informed the "New Failure of Nerve." In an article titled "The Heard-Huxley Paradise," published in the *Partisan Review*, Richard Chase attacked the British writer Aldous Huxley and his friend Gerald Heard, accusing them of using the language of mysticism to create a new form of "emotional totalitarianism."[115] In their attempt to escape the material world through an act of self-awareness, Huxley and Heard, Chase suspected, had resorted to draconian means to achieve this end. He described how they had applied a range of mystical techniques, prime among them being the strict curtailment of sex, in order to "supply the energy required for the enlargement of consciousness."[116] In Heard and Huxley's paradise, Chase recognized that in order to generate the energy needed to surpass the material boundaries of individualism (and enter a transcendental state) energy had to be placed under ever-tighter governance. Huxley, Chase recognized, had even conjured up a band of "psychic supermen called 'theocentric saints' or 'neo-Brahmins'" to run this utopia in a similar manner to a technocratic directorate.[117] Rather than "thinking" or "feeling" a new society into being, Chase observed, "they are willing to do almost any sort of intellectual violence to attain it."[118] "Even when they use the familiar terms of social science," "the 'tone' which we have come to associate with them has been dissipated and forgotten."[119] Their utopia, Chase believed, "is satisfied by diffusing the drives of the libido into a mystical love for society and also into what amounts to an infatuation with the universe."[120] Rather than carving out a utopia for an expanded consciousness, they had conjured up a police state where energy was regulated under the guise of mysticism.

The charge that Huxley used mysticism to conceal the rhetoric of scientific planning might have seemed apt given his satire of scientific technocratism in *Brave New World* (1932). However, as critics such as David Bradshaw have demonstrated, during the early 1930s Aldous Huxley had followed his brother Julian in recognizing the benefits of social planning, making Chase's accusations less far-fetched than they first appeared.[121] What is more, in the book that Chase criticized, a political work entitled *Ends and Means* (1937), Huxley provided a blueprint to achieve this society founded on principles of nonattachment, which depended, ironically, on placing rigid constraints on energy. The book was written following his emigration to the United States, where he spent some time at the late D. H. Lawrence's ranch in Taos, New Mexico, later settling in Hollywood, and finally moving to a ranch in Llano Del Rio, where he acted out the political necessity of decentralization by living off-grid with a small power generator. With Huxley becoming interested in mysticism through his involvement in the international peace movement, spiritual enlightenment served for him a similar purpose to planning: it would enable a wholesale redirection of energies. In this sense, he had developed a similar understanding of religion to that of his brother, grounding it in science, as he absorbed an expansive view of biology as a tool to explain all aspects of social life, from spiritualism to war, religion, politics, sex, and love.[122] Aldous Huxley's interest in mysticism therefore had little to do with organized religion but appealed as a state of being, a form of heightened awareness and intensified perception, leading him eventually to his experiments with hallucinogenic drugs in the mid-1950s. In *Ends and Means*, Huxley readdressed the positive and negative aspects of planning in light of this new spiritual outlook.

In the book Huxley described how his ideal utopia would be founded on the Brahman principle of "non-attachment" through disconnecting human instincts, "bodily sensations and lusts," "craving for power and possessions," "objects of these various desires," and "wealth, fame, social position."[123] The subject could reach this state only through adapting and controlling "thought-habits and action-habits," deploying a range of techniques, including fasting, yoga, continence, and meditation.[124] For a blueprint of this utopia, Huxley drew on *Sex and Culture* (1934), a book written by the British anthropologist J. D. Unwin.[125] Unwin had taken the antipathy between the libido and civilization (outlined in Freud's *Civilization and Its Discontents*) to heart and had posited a correlation between levels of social energy and the total amount of sexual energy used in a number of societies. He laid this out in *Sex and Culture*, where he documented the relationship between methods of continence used in a society and its correlate level of human energy. Needless to say, Unwin's theory was unwieldy, as he provided a vast chronicle of different societies in over four hundred pages, charting their sexual habits alongside their social practices and analyzing the different ways in which social energy was transformed into culture.[126] In the book Unwin defined four types of human societies: the zoistic, manistic, deistic, and rationalistic.[127] These levels were graded by the quantity of social energy displayed, which showed an inverse correlation to the amount of sexual energy spent. Rationalistic societies, which had the highest degree of pre- and postnuptial arrangements, demonstrated the greatest level of chastity and as a result had the highest amount of mental and social energy.[128]

This had a practical application within Huxley's utopia. "When high levels of sexual activity abounded," Huxley explained (paraphrasing Unwin), "societies lack energy and individuals are condemned to perpetual unawareness, attachment and animality."[129] For the individual to achieve "any kind of moral life superior to that of the animal," energy had to be channeled away from sex toward the disinterested ends of enlightenment.[130] After a period during which this productive energy was spent on the means of expansion, this social energy would turn toward the higher arts, as well as toward science, speculation, and social reform. This would lead to an elevated state that Unwin called (rather counterintuitively) "human entropy," to denote the moment when energy was driven toward greater refinement and accuracy. Just as the law of entropy inscribed direction to the second law, "human entropy" revealed "the Direction of the Cultural Process."

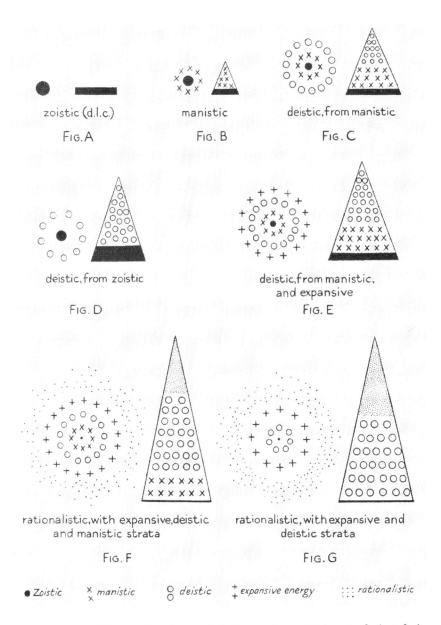

Diagrams of the different cultural states. J. D. Unwin, *Sex and Culture* (Oxford: Oxford University Press, 1934), 426.

It worked, Unwin explained, in that as far as a culture continues to generate energy all social behavior begins to work in a particular "Cultural Direction."[131] "Human entropy" could be maintained only once society remained chaste, and as previous civilizations had demonstrated, it would disintegrate as it reverted toward sexual freedom. It was this dissipation, Aldous Huxley believed, that had already begun to occur within the "liberal" community, whose members, by demanding freedoms of all sort, were abolishing "compulsory continence," "living as they do upon the capital of energy accumulated by a previous generation of monogamists, whose wives came to them as *virgines intactae*."[132] All around him, Huxley witnessed a reckless dissipation of energy that would lead to social breakdown, returning society to a barbaric state of animal attachment.

When released from sex, social energy was free to act as the "motive power" driving the expansive range of political, economic, educational, religious, and philosophical devices necessary within Huxley's utopia. And yet, the freeing up of social energy did not de facto ensure greater spiritual awareness, leading to a state of nonattachment. Instead returning to the imagery of the hydraulic ram, Huxley clarified that "mental and social energy" worked in the same way as the "energy of falling water; it can be used for any purpose that men choose to put it to—for bullying the weak and exploiting the poor just as well as for exploring the secrets of nature."[133] In fact, Huxley went on, the energy generated by chastity led more often to evil than to good—a principle demonstrated aptly by Hitler and Mussolini, who used similar mechanisms of forced chastity to harness energy into the channels of aggressive imperialism.[134] Having provided his own blueprint in *Ends and Means* for how to harness energy into greater ethical value, he concluded that "the particular problem of moralizing the energy produced by continence is the same as the general problem of realizing ideal ends."[135]

A few years later, Huxley reiterated the importance of these mechanisms of energetic control when he penned the introduction to Unwin's posthumous book *Hopousia; or, The Sexual and Economic Foundations of a New Society* (1940).[136] In the book Unwin outlined a sexual and economic blueprint that would enable society to display maximum energy. This was an elaborate plan including a proposed system of Alpha and Beta marriage arrangements and an economic system that eliminated usury, exchange value, and totalistic forms of political power.[137] Although agreeing with Unwin's central objective (the sensible organization of energy), Huxley pointed to its principal omission: it simplified the laws of causation. In focusing solely on the problem of energy, Unwin, Aldous Huxley recognized, had failed to account for the biological, technological, psychological, and intellectual elements that formed society. This was particularly important when it came to the problem of ethics; as Huxley pointed out, "Unwin's failure to describe the theology of Hopousia is a failure to describe a vital part of the mechanism for directing the energies of the Hopousians towards desirable ends."[138] Gesturing once again to the problem that more energy does not equate to social good, he returned to the disparity between quantity and quality: "a society may display immense energy," he concluded, "but if its energy is not directed by a philosophy based on all the facts of immediate, culturally unmodified human experience, it is likely to be at least as destructive as constructive, at least as evil as good."[139] Unwin's principal omission, according to Huxley, was that he failed to describe the theological armature through which his society would guide its energy—he had failed to supply the "ethically reputable channels" in which this energy would flow.[140] Ethics, for Huxley, operated as a hydraulic ram channeling energy toward the direction necessary to escape animal attachment and reach a state of expanded consciousness. It provided the "ethically reputable channels" that would guide a society's energy.

Julian Huxley's Energetic Ethics

The call for an ethical framework to channel energy to higher purposes expanded beyond those calling for the "rebirth of man" and the expansion of mysticism in the modern world. A parallel discourse framed energetic ethics in more technocratic terms as the ideal tool to remold the energetic basis of society and the political forms of the postwar world. Whereas Aldous Huxley maintained that the condition of nonattachment could be achieved by harnessing energy toward higher spiritual objectives, his brother Julian adopted a similar method of energetic control but deployed it to a contrary end in the construction of the ultimate planned structure: a world state.

Julian Huxley had long supported large-scale efforts in social planning, using his training as a biologist to create a holistic science through which all spheres of society—culture, religion, and politics—could be understood within a common medium. He found this in the language of psychoanalysis, which allowed the social world to be understood biologically as an energy system that required careful regulation.[141] For Julian Huxley, the Soviet Union (which he visited in 1931) was the prime example of this rational society, and he was largely impressed by their efforts in taking the social sphere under control.[142] Following his trip, he became a founding member of the planning group PEP that emerged following the publication of the "National Plan for Great Britain" by the *Week-End Review*, and he was a vocal advocate of the New Deal, having visited the Tennessee Valley Authority (TVA) twice (in 1935 and 1941), seeing it as the best example of social democracy in action.[143] On his appointment as director general to UNESCO in 1946, Julian Huxley understood the organization as an extension of these projects—an instrument to unify the direction of the world's energy through the aid of a secular religion he designated "scientific humanism."

Julian Huxley was not alone in seeing the political ideal of the "World State" as emerging from the will of the individual subject. Norman Cousins, after all, had done much to popularize this idea through his editorial. Lewis Mumford had similarly oriented the transformation of the human toward the goal of "One Worldism," arguing that this revolution could only come about by creating "unity in diversity" at a moment when all men, "integrated in all their functions, will have the vital energy to take part in this drama."[144] Long before his involvement with UNESCO, Huxley had called for "scientific humanism" to realize the ideal of world unity. Despite the period after the Second World War being a golden age for internationalism, Huxley, like many of his contemporaries (including his earlier collaborator H. G. Wells), had been advocating the political ideal of "One World" ever since the failure of the League of Nations and its associated program, the International Committee on Intellectual Cooperation.[145] Years prior to UNESCO becoming a reality, therefore, Huxley had popularized his brand of "scientific humanism" as a method to create a new world consciousness through a prodigal output of magazine articles, books, and radio broadcasts.[146]

Julian Huxley turned to the issue of founding a new religion based on science after reading Lord Morley's dictum that "the next great task of science will be to create a religion for humanity."[147] This religion, which Huxley designated interchangeably as "scientific humanism" and "evolutionary humanism," served a critical function: it provided a unified form for thought and action to replace a fractured idea system.[148] Having long been interested in Freudian psychoanalysis, Huxley's "scientific humanism" depended on establishing the unity between mind and matter and directing that system toward evolutionary ends. To do this, Huxley turned to energy as a unifying function that could reduce society, ethics, and religion into the same biological continuum. He had pushed this continuity farthest in *The Science of Life* (1931), a popular book written in collaboration with the science fiction writer H. G. Wells and his son George Philip Wells. Having recently finished *The Outline of History* (1926), Wells invited Huxley to collaborate on a new synthesis, focusing on the natural world and man's place within it. The book opened by describing man as an energy machine whose "energy-output" was directly aligned with their "energy-input" ("a mouse or a man works in much the same way as a petrol motor").[149] From this starting point, the trio concluded that "there is no observable life-process which is independent of its supply of physical energy," and as a result, they turned to the new science of ecology to explain how human society was continuous with nature's economy.[150] In the book they reduced all social relations to energy, describing how the "economy of nature" could be managed as a unit controlled in the same way as "a board of directors plans a business."[151] Taking the world as one complete unit, "from the standpoint of biological economics," the book argued that the responsibility of human stewardship was "to make the vital circulation of matter and energy as swift, efficient, and wasteless as it can be made."[152]

When Julian Huxley extended this outlook toward religion, as he had done in his previous book, *Religion without Revelation* (1927), he saw it in comparable terms, as a mechanism to ensure the efficient regulation of energy. By applying modern psychology to excavate religion, he came to the conclusion, similar to that of William James, that "revelation" undergone in religious experience was "revelation only in the psychological sense, not literally."[153] Revelation did not depend on an impersonal God but was a mental state like any other, with its own method of libidinal organization.[154] The anthropomorphism of God, Huxley described, operated as "organisations of human thought which seek to represent, canalise, and give a comprehensible interpretation of the forces affecting human destiny."[155] Following this logic, "the basic postulate of evolutionary humanism," Huxley explained, was "that mental and spiritual forces—using the term *force* in a loose and general sense—do have operative effect, and are indeed of decisive importance in the highly practical business of working out human destiny."[156] These forces, Huxley believed, could be divided between good and evil, "hate, envy, despair," and "love in the broadest sense."[157] Through the "constructive disposition of force," inner harmony could be brought about both in the individual and in the larger social system. This was because, Huxley clarified, forces

> operate not only within individual minds, but through the social framework. A society may be so organised that it generates large amounts of hate or envy or despair; or creates vast tracts of ugliness; or imposes subnormal health or inadequate mental development on large sections of its population. Or its organisation may serve to encourage and facilitate constructive enthusiasm, to create beauty, and to promote full and healthy individual development.... The fact remains that social organisation does canalise and concentrate the psychological forces of human nature in different ways, so that society can act either as an organ of frustration or an organ of fulfilment.[158]

Huxley gave an example of his scientific religion in action in 1934, when he provided a "philosophy of dictatorship" in his book *If I Were Dictator* (1934). In this book, written for a series on the subject, Huxley set himself against a society in which the citizen is "so occupied with the business of getting a living that they have little time or energy for life; most of their rich potentials are unrealised."[159] To correct this misdirected energy, Huxley set about proposing new institutions that would enable a healthy expenditure of energy. The constraints of "herd instinct, history, common language" brought about by nationalism operated as a constraining energy pattern, enabling "a surprisingly strong inner driving force, which for the moment at least is easily able to override the newer driving force towards internationalism and world organization."[160] To counter this, Huxley would impose both practical and ideal methods to replace archaic structures with ones that worked toward the rational goals of economic planning and internationalism. "Scientific humanism," he explained, would provide "the vision of the remote goals of this continuing change," enabling "outlets for the desire that overflows the Here and Now—an ideal, but a practical ideal, so that its energies are focused on this life, not on another, nor dissipated in some nebulous unreality."[161]

By 1942, as war raged around him, Julian Huxley's commitment to "scientific humanism" became even stronger. His essay "On Living in a Revolution," published in *Harper's* (and included in a collection of essays under the same title), described how the revolution suggested in *If I Were Dictator* was already in progress. Caused by the war and the collapse of the laissez-faire system and the sovereign nation-state, the revolution had enforced a new social architecture: greater planning, a subordination of economic to social needs, a higher degree of social integration, and a move toward international organization.[162] Although Huxley recognized these systems to "already be in revolution," the patterns were not fixed, and he feared totalitarianism to be just as likely an outcome as democracy.[163] To prevent a slip into the former, Huxley argued for a more suitable social "yardstick," adopting the "American expression" that had played such an ideological role in the foundation of the TVA.[164] Yet while the yardstick discussed in the TVA appealed as an objective standard, Huxley would relocate it within the subject, who would become a register of the level of democratic participation. In this model "the yardstick by which we can measure democratic achievement" would, Huxley believed, represent "the satisfaction of the needs of human individuals, and the yardstick by which we can measure democratic method is their active and voluntary participation in all kinds of activities."[165] Huxley reiterated the importance of this yardstick metaphor in his high-profile Romanes Lecture the next year. There he called for the individual to become a social yardstick to bring about a new political constitution. Once again, the desirable direction of evolution would provide the most comprehensive yardstick against which the ethical system could be modeled.[166] With ethics operating like a physical network, differentiating natural energy loads from unnatural ones, the social "yardstick" would be judged by the capacity to direct energy toward progressive ends, which in Huxley's framework would be toward world unity.[167]

Redirecting the World's Energy

On becoming the director general of UNESCO in 1946, Julian Huxley viewed the organization as an extension of his own scientific program.[168] As such, he quickly overreached his diplomatic position, promoting the organization as a practical organ for his "evolutionary humanism"—a means to restructure the world's energy. It is hardly surprising that energy would become a key metaphor in understanding how a "world consciousness" might unify the world's racial and cultural diversity. Energy had long driven the conceptual frameworks to imagine the political ideal of world unity. In fact, three of the most prominent figures to promote internationalism in the interwar period, Henri Bergson, Jan Christiaan Smuts, and Albert Einstein, each had a unique conceptual understanding of energy at the core of their political internationalism.[169] Furthermore, it is not hard to see why energy lent itself so neatly to the political ideal of world unity. The distinctive qualities of energy—its monism, its universality, and the impossibility of containing it within arbitrary boundaries (such as the nation-state)—made it particularly well suited to imaginaries of a continuous international world order.

By 1927, for instance, the vision of the globe as an integrated organism put forward by the Russian biophysicist Vladimir Vernadsky had provided a framework within which to understand the earth as a single system united by the free circulation of energy from the sun.[170] Once the earth could be understood as a complete energy system, it drove conceptual paradigms of a new political order. The growing awareness of the earth as a single organism, for instance, allowed H. G. Wells, the most vocal political advocate of internationalism during this period, to imagine a new set of political and economic mechanisms suited for a global abode.[171] Wells had put forward an early model of this in *A Modern Utopia* (1905), where he had gone so far as to fantasize about an economic model of "bookkeeping" within the "World State," whereby everything sold on the market could be accounted for in terms of "units of physical energy" carried by the finite planet at any one time.[172] This continued to saturate Wells's writing during the 1920s and 1930s as he drew on the emerging language of ecology to imagine an integrated world. In *The Science of Life* both he and Julian Huxley also stressed the importance of regulating food and energy "as one community, on a worldwide basis; and as a species."[173] The paradigm of the earth as a united energy system thus lent itself to discussions of efficient global management, making the political structure of the nation-state appear to be an anachronistic political form undermined by energy laws.[174] Wilhelm Ostwald's energetics, after all, had led to a similar conclusion, as he combined German imperialism with his understanding of energy to envision a monist civilization—a world united through the efficient organization of energy.

Drawing on this intellectual lineage, Julian Huxley developed his own "purpose and philosophy" of the organization in a short book published in 1946 under the title *UNESCO: Its Purpose and Its Philosophy*. Written without the permission of the executive body and denied official UN status, Huxley's text offered further discussion of his scientific humanism, situating UNESCO as the tool to achieve this end.[175] In the body of the text, Huxley rehearsed the central tenets of scientific humanism and suggested a range of practical programs that could be utilized to divert the expenditure of energy away from nationalistic pursuits, toward the creation of a "world mind."[176] In the pamphlet Huxley suppressed the overtly Freudian armature that informed his other work, although it is clear that he conceptualized UNESCO as a natural elaboration of his earlier philosophy. He reiterated arguments put forward in countless other places that maintained that the foremost role of education in the modern world was to redeploy a quantity of "mental energy" in order to minimize conflict and to elaborate a world consciousness—that is, to "modify this primitive psychical morphology into something less wasteful for the purposes of adult human existence."[177] This took on a special resonance as he came to discuss the purpose of UNESCO, as he posed a range of practical programs to achieve "the emergence of a single world culture."[178] For Huxley, "unification in the things of the mind" by developing a "common pool of experience, awareness and purpose" would "lay the foundations on which world political unity can later be built."[179] Through UNESCO's endeavors in cultural exchange, international understanding, and cooperation, it is not hard to see how Huxley might have envisioned the organization as the ideal tool to concentrate the world's energy.

When read through the lens of his "evolutionary humanism," it becomes apparent that Huxley understood the organization to be a synthetic infrastructure through which to focus the world's energy around a unified culture and belief system. In this respect, he shared much with the French Jesuit priest Pierre Teilhard de Chardin. Huxley had met Teilhard in Paris in 1946 while working at the organization, and he was amazed to find someone who had been on exactly the same course of thought for over three decades.[180] Almost a decade later, writing the introduction to the first English translation of Teilhard's *The Phenomenon of Man*, Julian Huxley reflected on how Teilhard's paradigm of the noosphere was a "parallel" version of his scientific humanism, which had been in development since 1913.[181] Teilhard had developed the concept of the noosphere, along with the Russian biophysicist Vladimir Vernadsky, to describe a sphere of human consciousness and thought that rested on top of the biosphere.[182] As humans were the only group that thought together, it was psychic energy, Teilhard believed, that led the evolutionary process forward.[183] This climaxed in a "point Omega" that, driven by a "radial energy," defied the law of entropy to unite all thought in a central point that enclosed the world. In Teilhard's vision of the noosphere could be found the powerful vision of unity promised by UNESCO, where the individual extended toward a "global, scientific, evolutionary humanism."[184] Huxley described how Teilhard's vision of a radial energy provided an "energetic tension" in which "idea will encounter idea, and the result will be an organized web of thought, a noetic system operating under high tension, a piece of evolutionary machinery capable of generating high psychosocial energy."[185]

It is important to note that just as energy was put to work as a unifier, it also operated as a disciplinary tool to conceal national and racial difference. The image of the globe as an integrated energy system (still two decades away from being immortalized by the "Earthrise" photo taken in 1968 during the Apollo 8 mission) reduced the diversity of all cultures to a monolithic entity—encapsulated by the most universal of substances: energy. This reductionist worldview was hardly novel, and Glenda Sluga has pointed out that Huxley's vision of a "world consciousness" (and much of the discourse that dominated postwar cosmopolitan internationalism) reiterated the imperialist language of empire building, continuing to posit the "white race" as a civilizing agent in this process.[186] As Huxley made explicit within his text, it was the responsibility of the "civilizing nations" to steward the world's energy, to concentrate it under a series of "universal" Western values.

Nevertheless, even though the organization distanced itself from Huxley's manifesto, Huxley's vision of UNESCO as the "world-mind" stuck, and the text circulated widely. This was not helped by the publisher, which, despite including the disclaimer on the dust jacket that Huxley's text did not "reflect the official policies of the commission," promised that "no one can speak with greater authority about the purpose and philosophy of UNESCO than Julian Huxley."[187] As Huxley framed UNESCO as an extension of his scientific humanism, he demonstrated the conflux of social planning and religion on a global scale. To him, the organization would become a synthetic infrastructure through which to focus the world's energies toward a nodal point. Here the long-held dream of concentration, of focus, of center, was used to imagine one of the foremost political organs of the postwar world: the world's energy united in direction through one supranational "scientific religion."

Conclusion

Energetic ethics came to define how energy was understood in the early years of the atomic age. Far from reflecting a new atomic consciousness, energy continued to be understood through the same paradigms that had developed over the first decades of the twentieth century: conceptualized as an aqueous fluid that had to be managed and harnessed through society. The flexibility of this discourse allowed it to expand across the political spectrum, defining a range of different intellectual programs, from a broader cultural debate about the "new man," to providing a framework through which to understand the power of totalitarianism, to a defense of technocratic internationalism. As New Dealers came to terms with an entirely new political landscape in the years leading up to and following the Second World War, factions on the political left began to replace the physical infrastructure of the watercourse with an ideational one, whose channels would be found within the individual psyche.

While the conceptual frameworks used to understand energy remained relatively unchanged by the postwar period, however, energy's relationship to moral value had shifted. Whereas in the early 1930s energy had been valued as an objective "yardstick" for society, as the decade progressed it could no longer be opposed to social form. Once the "psychedelic revolution" kicked off in the late fifties, initiated in 1953 by Aldous Huxley's ingestion of mescaline crystals, the physical "gateway" became a chemical one, the flow of energy controlled as much by an acid pill as by the placement of a hydroelectric dam on a physical waterway.[188] The description of Huxley's first acid trip, captured in vivid detail in *The Doors of Perception* (1954), hardly differed from his early method of forced chastity or his older brother's blueprint for "One World."[189] Both depended on a planned intervention in the flow of energy: for Aldous Huxley this meant a radical expansion of consciousness, whereas for Julian Huxley it took the form of concentration toward world unity. As the next generation collapsed spiritual enlightenment with drug taking, they did not lose the earlier infatuation with controlling energy. Rather, as they came to focus on expanding the boundaries of consciousness, they simply transformed it to another plane.

Thus, in an ironic twist, on opening *Playboy* magazine in 1967, among titillating breasts and erotic thighs, one could find Julian Huxley, then eighty years old, lauding "psychedelic drugs" (along with Yoga and Zen Buddhism) as a constructive social tool to aid his objective of "evolutionary humanism."[190] With Aldous now dead, Julian Huxley located himself amid the social revolution of the swinging sixties, characterized by the strains of overpopulation, the Soviet Bloc, growing urban sprawl, pollution, and the onset of computerized automation. Despite this unfamiliar world, Huxley repeated an old adage unchanged for three decades: the destiny of mankind depended on its ability "to provide truly satisfying goals for human beings everywhere, so as to energize our species, to stimulate it to move and to ensure that it moves in the right direction."[191] Once again, this would be a two-way affair, depending on the foundation of world unity and a global ecology, while also focusing on the evolutionary advancement of the subject, "planning for greater fulfilment for human individuals."[192] In this way, as Julian Huxley praised psychedelic drugs for being (alongside the "World State") an effective infrastructure through which to redirect the flow of energy, the external yardstick moved inward, with the autonomous subject taking on the political promise energy had once revealed. Rather than energy representing an "external yardstick" against which to measure society, the world's energy system would depend on an internal yardstick, its entire system revolutionized by a tiny acid pill.

Epilogue

MEOW

> energy
> anyway—in a society like America energy if it is not moral is only
> material. Which cannot be destroyed is never destroyed is only
> left all over the place. Junk.
> —*Charles Olson, "The Ocean," 1965[1]*

During the bitterly cold winter of 1977, the newly elected president Jimmy Carter sat by a roaring fire, wrapped in a cardigan, and asked the nation to follow his example, to turn their thermostats down to 65 degrees to conserve the nation's energy.[2] By July 15, 1979, as the energy crisis worsened and Carter's approval ratings plummeted as a result of rapid inflation and soaring gasoline prices, Carter made a leap of faith and tied the energy crisis to the condition of national character. With Americans having been unmoved by his previous four energy speeches, about which one journalist claimed he assumed the role of an "evangelist, preaching to a nation living in energy sin," Carter delivered another address to the nation on the energy crisis, hoping this time to garner the momentum necessary to implement a solid conservation plan.[3] During his speech, officially titled "Energy and National Goals" (but nicknamed the "Malaise Speech"), Carter conveyed a hard message to the one hundred million viewers watching from their living rooms. The nation was facing a problem "deeper than gasoline lines or energy shortages, deeper even than inflation or recession."[4] It was suffering from a "crisis of confidence" that "strikes at the very heart and soul and spirit of our national will."[5] Drawing on the recent critiques of consumer affluence put forward by Daniel Bell, Christopher Lasch, and Robert Bellah, Carter, a born-again Christian, charged the nation with wanton materialism leading to greed, lack of purpose, and the dissolution of faith in higher values.[6] To overcome this current discontent, Carter stressed the problem of energy, hoping that the nation, by focusing on this issue, could reignite its core values. Energy, he promised, would be "the immediate test of our ability to unite the Nation, and it can also be the standard around which we rally. On the battlefield of energy we can win for our Nation a new confidence, and we can seize control again of our common destiny."[7] By taking control of energy through carpooling, using public transport, obeying the speed limit, and turning the thermostat down, the nation, Carter trusted, could achieve the levels of unity needed to overcome the vacuous consumer culture pervading society.

The stress on energy conservation that lay at the center of Carter's administration has often been read as the origin of a new energy consciousness that emerged from a decade of fuel shortages, price hikes, and doomsday scenarios, brought about by the dominance of the Organization of the Petroleum Exporting Countries, unrest in the Middle East, and a growing environmental movement.[8] And yet, as one reporter from the *St. Louis Post-Dispatch* recognized, with Carter having "mixed morality with energy," it was not hard to see how his speech sat within a longer tradition of moralizing energy.[9] In fact, during his speech, Carter had drawn on William James's "moral equivalent" to capture the magnitude of the challenge the public faced, urging the nation to find in the energy crisis a "moral equivalent of war."[10] During his address, not only would Carter tell of one labor leader who begged him, "When we enter the moral equivalent of war, Mr. President, don't issue us BB guns," but he also made it clear, as he described the "energy battle" to be won in both material and spiritual terms, that the war on energy was going to be fought on both fronts. Just as Carter begged the public to overhaul their consumption habits, so too did he hope that by fighting a war on the "energy battlefield" the nation would be able to redirect its spiritual energy away from conspicuous consumption toward a higher purpose and new values.

Even though Carter's critics mocked the emotion that his "moral equivalent" conjured up, reducing it to its unfortunate acronym MEOW, Carter's nod toward James illuminates how the problem of energy had long circulated within American thought. As the nation's energy system broke down, leading to queues at gasoline pumps and freezing apartments, the material blockage was symbolic of a larger metaphorical threat that had long shaped the public imaginary. As this book has demonstrated, its roots could be found in the early decades of the twentieth century. It was located in Brooks Adams's "Law," with its neurosis that economic man was dissipating energy through "Greed" rather than conserving it in "Fear"; in concerns over misdirected energy, as immigrant communities were characterized as lacking the adequate controls on their energy; and in broader anxieties that American vigor would drain away with a warming climate.

The Regional Planning Association of America had also drawn on it, as it envisioned redirecting the channels of energy flow away from the congested urban centers and back into the regions. It had informed New Deal rhetoric, as figures such as Stuart Chase and Howard Scott surveyed the abundance of energy in goods going to waste in factories—a symbol of how energy had been purposefully withheld from the public by the manipulative controls of the price system. For others, such as Julian Huxley, schooled in Freudian psychoanalysis, it encapsulated an unhealthy state of repression as society harnessed its energy toward laissez-faire capitalism and pernicious forms of nationalism rather than toward world unity. In contrast, free market advocates, such as Isabel Paterson, located the blockage (or the "short-circuit of energy") in the role of the state, maintaining that for energy to circulate freely it required a "god of the machine" liberated from the artificial controls of government. Far from emerging out of a seventies "malaise," therefore, the problem of how energy flowed through society had long been a subject of concern, sitting at the heart of public discourse about the condition of American society—both its symbolic forms and its social institutions.

Throughout this period, energy remained a moral problem. At the heart of this was the question of energy's relationship to society. Was society a measure of its energy? Or, on the contrary, should society be valued by its ability to control and manage its energy toward higher ends? This question inevitably led to a range of conceptual problems. Was energy external to society, or could it only be harnessed through it? Was there an intrinsic relationship between energy and social good? How could qualitative states such as thinking and writing be reduced to energy? And what was energy's value? Information theory, emerging between 1943 and 1954 out of the Macy Conferences held in New York under the auspices of the Josiah Macy Jr. Foundation, would provide answers to some of these questions as it shifted the emphasis away from energy and matter and toward information as an explanatory device. This change clarified many of the confusions that had surrounded earlier discussions of energy. For instance, the problem of how to quantify the amount of energy units utilized within intellectual labor by the writer, lawyer, or politician could now be explained by way of information. Not only did information theory clarify some of these questions, but it made them increasingly obsolete, as the world began to be understood in terms of communication. As a result, the questions that were not solved by this new interdisciplinary field, such as the connection between energy and social good, were left unanswered as information became the dominant mode of understanding human systems. As Norbert Wiener, one of the founders of cybernetics, famously concluded, having considered the imbalance shown by the "vanishingly small" proportion of energy used in the computing machine (and the human brain) in relation to the remarkable "performance of the apparatus," "information is information, not matter or energy. No materialism which does not admit this can survive at the present day."[11]

And yet, although the emergence of information theory supplied an explanation for some of the earlier problems, the energy language that informed Carter's speech illustrates how by the 1970s energy had not escaped these debates.[12] Energy continued to be a cultural trope through which to envisage the health of the nation. Far from being undermined by novel scientific paradigms, energy remained as ethical, metaphorical, and rhetorically loaded as it had been at the start of the century. Carter, moreover, recognized the power of energy to grip the national psyche. Energy, as Carter accepted, had a far more potent conceptual life when it was abstracted from its material life, no longer reduced to the scientific principle of energy conservation or scarce resources. For it to *move* the nation, it had to be a moral problem.

This moral energy did not disappear after the fuel crisis; it continues to inform the public imaginary today. As societies confront the threat of anthropogenic climate change, moral discourse is increasingly being used to effect behavioral change and reduce energy consumption. But even though behaviors associated with fighting climate change have become highly moral, it remains hard to think of energy itself as connected to ethical frameworks. If anything, the excessive burning of fossil fuels over the past century is increasingly being viewed as the greatest moral failing of our time.

There is a disconnect, therefore, between our relationship with fossil fuels and how we think about energy, which remains highly moralized. Far from being limited to scientific or technical paradigms, energy remains as abstract and metaphysical as it was at the turn of the century. These metaphors have become so completely assimilated into discourse—so commonsense—that we no longer recognize the hold they have over the social imaginary. For example, as Cara New Daggett has shown, the ethical framework of late capitalism, with its "commitment to growth and productivity," "has been reinforced by a geo-theology of energy" that ties productive work to energy.[13] These conceptual frameworks remain far from academic propositions. Instead, they are deeply rooted in our social structures, economic models, industrial policies, and technological systems. As Nicola Labanca has demonstrated, energy metaphors have fundamentally shaped technical infrastructures and policy decisions in the energy sector.[14] Jenny Rinkinen, Elizabeth Shove, and Jacopo Torriti have also exposed the range of "energy fables" that dominate contemporary energy speak, showing how phrases such as "energy demand," "efficiency," and the "energy trilemma" are based on confused metaphors. Far from being limited to folklore, these "energy fables" have become central in framing future energy challenges and technical solutions within the energy sector.[15]

As increased attention is paid to the material effects of these "energy fables," there is a growing awareness that we need to uncover—and expose—the "stories" embedded in our current energy frameworks. Daggett, for example, argues that only once we excavate the ethical roots of our dependence on energy can we start to reimagine these dominant energy tropes and "multiply energy epistemologies, metaphors, and visions concerning how we participate in and value work, production, and dynamism."[16] Running through this body of thought is a growing acknowledgment that we need to revise, rework, and make explicit a new ethics of energy. What has disappeared from view, therefore, is not moral energy itself but a broader public discourse that debates, exposes, and poses alternatives to the energy metaphors that have driven our carbon-intensive regimes.

By uncovering the roots of America's energy consciousness, this book has illuminated the United States' vigorous historical public discourse on energy. This public discourse not only reinforced but also remained attentive to the common assumptions embedded within our current energy metaphors—such as the relationship between force and form, the relationship between energy and social good, and energy's intrinsic value. These questions are even more relevant today than during the first half of the twentieth century. As the challenge of rapid decarbonization becomes increasingly urgent, societies will have to reflect on these questions as they redesign their technical systems and ways of living. Only once we recognize how our energy frameworks are rooted in a specific way of imagining and describing energy can we develop new institutions, social practices, infrastructures, and representations that will transform the stories we tell about energy and our relationship to it. Returning moral energy to public discourse, therefore, will provide new perspectives and ethical frameworks that describe our relationship to energy. If, as Carter recognized, technological or economic arguments rarely "move" the nation into action, reframing energy as a fundamentally moral question could become a unique tool in negotiating a new set of relationships between society and our energy systems during a time of global warming and environmental collapse.

NOTES

Introduction

1. J. Dewey, "Force and Coercion," *International Journal of Ethics* 26, no. 3 (April 1, 1916): 361.

2. Dewey put this argument forward in a series of high-profile essays. J. Dewey, "Force, Violence and Law," *New Republic* 5, no. 64 (1916): 295–297; Dewey, "Force and Coercion," 361. This debate has played a central role in American intellectual history, incorporated into narratives of the intellectual left, the decline of progressivism, and pragmatism. For more on this debate, see J. P. Diggins, *The Promise of Pragmatism: Modernism and the Crisis of Knowledge and Authority* (Chicago: University of Chicago Press, 1994), 250–259; C. N. Blake, *Beloved Community: The Cultural Criticism of Randolph Bourne, Van Wyck Brooks, Waldo Frank and Lewis Mumford* (Chapel Hill: University of North Carolina Press, 1990), 157–180; D. Levine, "Randolph Bourne, John Dewey and the Legacy of Liberalism," *Antioch Review* 29, no. 2 (1969): 234–244; J. Livingston, "War and the Intellectuals: Bourne, Dewey, and the Fate of Pragmatism," *Journal of the Gilded Age and Progressive Era* 2, no. 4 (2003): 431–450.

3. R. Bourne, "Twilight of Idols," *Seven Arts* 11 (1917): 688–702.

4. Bourne, "Twilight of Idols."

5. Critics have argued that Bourne misunderstood Dewey by overstressing his instrumentalism. R. B. Westbrook, *John Dewey and American Democracy* (Ithaca, NY: Cornell University Press, 1991), 197–212.

6. Bourne, "Twilight of Idols," 691.

7. Donald Trump, "Remarks by President Trump at the Unleashing American Energy Event" (speech, US Department of Energy, Washington, DC, June 29, 2017), https://web.archive.org/web/20210120195650/https://trumpwhitehouse.archives.gov/briefings-statements/remarks-president-trump-unleashing-american-energy-event/.

8. B. Johnson, *Carbon Nation: Fossil Fuels in the Making of American Culture* (Lawrence: University Press of Kansas, 2014), xxvii–xxix.

9. Johnson, *Carbon Nation*, xxi. Scholars have often remarked on the invisibility of fossil fuels within cultural production, a claim first made by Amitav Ghosh in 1992 in his review "Petrofiction: The Oil Encounter and the Novel," *New Republic* 206, no. 9 (1992): 29–34. Increasingly this absence has been reinterpreted as an omnipresence, or, to use Patricia Yaeger's phrase, an "energy unconscious" whereby, even if not directly

referenced, fossil fuels are inscribed at all levels of a text. See P. Yaeger, "Editor's Column: Literature in the Ages of Wood, Tallow, Coal, Whale Oil, Gasoline, Atomic Power, and Other Energy Sources," *PMLA* 126, no. 2 (2011): 305–310. See also P. Hitchcock, "Oil in an American Imaginary," *New Formations* 69, no. 4 (2010): 81–97; G. Macdonald, "Research Note: The Resources of Fiction," *Reviews in Cultural Theory* 4, no. 2 (2013): 1–24.

10. The 1970s has often been cited as a transitional moment in America's relationship to fossil fuels, brought about by the oil embargo caused by the Arab-Israeli war in 1973–1974. As Caleb Wellum has shown, this crisis served many different social and political purposes in postwar US political culture and became a moment when Americans became increasingly self-aware and critically engaged with their dependence on fossil fuels. C. Wellum, *Energizing Neoliberalism: The 1970s Energy Crisis and the Making of Modern America* (Baltimore: Johns Hopkins University Press, 2023). See also Johnson, *Carbon Nation*, xviii.

11. Trump, "Remarks by President Trump."

12. "A Call for Greatness," *New York Times*, February 8, 1960, quoted in J. W. Jeffries, "The 'Quest for National Purpose' of 1960," *American Quarterly* 30, no. 4. (1978): 459.

13. J. Carter, "Energy and National Goals" (July 15, 1979), in *Public Papers of the Presidents of the United States, Jimmy Carter, 1979, Book II, June 23–December 31* (Washington, DC: US Government Printing Office, 1979), 1235–1241.

14. The growing field of energy history has largely focused on the materiality of energy transitions. Canonical works such as A. Crosby's *Children of the Sun: A History of Humanity's Unappeasable Appetite for Energy* (New York: W. W. Norton, 2006), V. Smil's *Energy in World History* (Boulder, CO: Westview, 1994), E. A. Wrigley's *Energy and the English Industrial Revolution* (Cambridge: Cambridge University Press, 2010), and A. Kander, P. Malanima, and P. Warde's *Power to the People: Energy in Europe over the Last Five Centuries* (Princeton, NJ: Princeton University Press, 2018) have focused on the material, economic, and technical development of energy systems. Increasingly, attention has been paid to the role of social and cultural factors in shaping these transitions. See D. Nye, *Consuming Power: A Social History of American Energies* (Cambridge, MA: MIT Press, 1998); D. Nye, *Electrifying America: Social Meanings of a New Technology, 1880–1940* (Cambridge, MA: MIT Press, 1990); A. Needham, *Power Lines: Phoenix and the Making of the Modern Southwest* (Princeton, NJ: Princeton University Press, 2015); G. Gooday, *Domesticating Electricity: Technology, Uncertainty, and Gender, 1880–1914* (London: Pickering & Chatto, 2008); C. F. Jones, *Routes of Power: Energy and Modern America* (Cambridge, MA: Harvard University Press, 2014); A. Harrison Moore and R. W. Sandwell, *In a New Light: Histories of Women and Energy* (Montreal: McGill–Queens University Press, 2021).

15. Scholarship in the energy humanities has grown rapidly in the past few years. For examples of the many works emerging in this area, see I. Szeman and D. Boyer, *Energy Humanities: An Anthology* (Baltimore: Johns Hopkins University Press, 2017); S. LeMenager, *Living Oil: Petroleum Culture in the American Century* (New York: Oxford University Press, 2016); Johnson, *Carbon Nation*; B. Johnson, *Mineral Rites: An Archaeology of the Fossil Economy* (Baltimore: Johns Hopkins University Press, 2019); R. Barrett and D. Worden, *Oil Culture* (Minneapolis: University of Minnesota Press, 2014); C. N. Daggett, *The Birth of Energy: Fossil Fuels, Thermodynamics, and the Politics of Work*

(Durham, NC: Duke University Press, 2019); M. T. Huber, *Lifeblood: Oil, Freedom, and the Forces of Capital* (Minneapolis: University of Minnesota Press, 2013); T. Mitchell, *Carbon Democracy: Political Power in the Age of Oil* (London: Verso, 2013); S. Wilson, A. Carlson, and I. Szeman, *Petrocultures: Oil, Politics, and Culture* (Montreal: McGill–Queen's University Press, 2017); K. Pinkus, *Fuel: A Speculative Dictionary* (Minneapolis: University of Minnesota Press, 2016); I. Szeman, J. Wenzel, and P. Yaeger, *Fueling Culture: 101 Words for Energy and Environment* (New York: Fordham University Press, 2017); H. C. M. Scott, *Fuel: An Ecocritical History* (London: Bloomsbury Academic, 2018); A. Goodbody and B. Smith, "Stories of Energy: Narrative in the Energy Humanities," *Resilience: A Journal of the Environmental Humanities* 6, nos. 2–3 (2019): 1–25.

16. LeMenager, *Living Oil*; B. Black, *Petrolia: The Landscape of America's First Oil Boom* (Baltimore: Johns Hopkins University Press, 2000); B. Black, "Oil for Living: Petroleum and American Conspicuous Consumption," *Journal of American History* 99, no. 1 (2012): 40–50; Johnson, *Mineral Rites*; Johnson, *Carbon Nation*; F. Buell, "A Short History of Oil Cultures: Or, the Marriage of Catastrophe and Exuberance," *Journal of American Studies* 46, no. 2 (2012): 273–293. A special issue edited by Ross Barrett and Daniel Worden of the *Journal on American Studies* 46, no. 2 (2012) on "Oil Cultures" also examines this subject; see also the special issue "Energy Pasts and Futures in American Studies," edited by Natasha Zaretsky, Michael Ziser, and Julie Sze, *American Quarterly* 72, no. 3 (2020).

17. Huber, *Lifeblood*; Mitchell, *Carbon Democracy*; Wellum, *Energizing Neoliberalism*.

18. Huber, *Lifeblood*, xv–xvi.

19. J. Lieberman, *Power Lines: Electricity in American Life and Letters, 1882–1952* (Cambridge: Cambridge University Press, 2017), 4. See also S. Halliday, *Science and Technology in the Age of Hawthorne, Melville, Twain, and James: Thinking and Writing Electricity* (New York: Palgrave Macmillan, 2007); C. T. de la Peña, *The Body Electric: How Strange Machines Built the Modern American* (New York: New York University Press, 2005); Nye, *Electrifying America*.

20. R. Wright and F. Trentmann, "The Social Life of Energy Futures: Experts, Consumers and Demand in the Golden Age of Modernization, c. 1900–1973," in *Economy and Environment in the Hand of Experts*, ed. F. Trentmann, A. B. Sum, and M. Rivera (Munich: Oekom, 2018), 60–61; Daggett, *Birth of Energy*, 4. See also R. Lane, "The Nature of Growth: The Postwar History of the Economy, Energy and the Environment" (PhD diss., University of Sussex, 2015).

21. S. Kern, *The Culture of Time and Space, 1880–1918* (Cambridge, MA: Harvard University Press, 2003), 8–9.

22. Szeman and Boyer, *Energy Humanities*, 3.

23. See B. Clarke, *Energy Forms: Allegory and Science in the Era of Classical Thermodynamics* (Ann Arbor: University of Michigan Press, 2001); B. Clarke and L. D. Henderson, *From Energy to Information: Representation in Science and Technology, Art, and Literature* (Stanford, CA: Stanford University Press, 2002); B. J. Gold, *ThermoPoetics: Energy in Victorian Literature and Science* (Cambridge, MA: MIT Press, 2010); A. MacDuffie, *Victorian Literature: Energy, and the Ecological Imagination* (Cambridge: Cambridge University Press, 2014); C. Smith, *The Science of Energy: A Cultural History of Energy Physics in Victorian Britain* (Chicago: University of Chicago Press, 1998); R. Martin, *American Literature and the Universe of Force* (Durham, NC: Duke University

Press, 1981); A. Rabinbach, *The Human Motor: Energy, Fatigue, and the Origins of Modernity* (Berkeley: University of California Press, 2016); T. Y. Choi, "Forms of Closure: The First Law of Thermodynamics and Victorian Narrative," *ELH* 74 (2007): 301–322. It has been well documented how early scientific understandings of energy emerged out of, and reflected, the major theological, ethical, and social questions that defined Victorian Britain. For more on this, see C. Smith, "Natural Philosophy and Thermodynamics: William Thomson and 'The Dynamical Theory of Heat,'" *British Journal for the History of Science* 9, no. 3 (1976): 293–319; G. Myers, "Nineteenth-Century Popularizations of Thermodynamics and the Rhetoric of Social Prophecy," *Victorian Studies* 29, no. 1 (1985): 35–66; G. Gooday, "Sunspots, Weather and the Unseen Universe: Balfour Stewart's Anti-materialist Representations of 'Energy' in British Periodicals," in *Science Serialized: Representations of the Sciences in Nineteenth-Century Periodicals*, ed. G. Cantor and S. Shuttleworth (Cambridge, MA: MIT Press, 2004), 111–147.

24. Daggett, *Birth of Energy*, 6.

25. Daggett, *Birth of Energy*, 3.

26. Rabinbach, *Human Motor*.

27. J. R. McNeill and P. Engelke, *The Great Acceleration: An Environmental History of the Anthropocene since 1945* (Cambridge, MA: Harvard University Press, 2014). This fossil fuel narrative is also rooted in American origin stories that extend back to early colonial visions of the New World as a "Garden of Eden," a land of abundance to be exploited. See C. Merchant, *Reinventing Eden: The Fate of Nature in Western Culture* (London: Routledge, 2013).

28. Johnson, *Carbon Nation*, 40.

29. See, e.g., Jones, *Routes of Power*; Nye, *Consuming Power*; T. P. Hughes, *Networks of Power: Electrification in Western Society, 1880–1930* (Baltimore: Johns Hopkins University Press, 1983).

30. As noted in note 15, there is a growing body of scholarship that examines cultural responses to the emergence of energy systems in the United States. These, however, have largely remained focused on fossil fuels and their impact on American culture and letters. Less attention has been paid to the intellectual debates and concerns that were occurring adjacent to the fossil economy.

31. Daggett has stressed the importance of decoupling energy from work. Daggett, *Birth of Energy*, 11.

32. Daggett, *Birth of Energy*.

33. P. Crutzen and E. F Stoermer, "The Anthropocene," *Global Change Newsletter* 41 (2000): 17–18.

34. McNeill and Engelke, *Great Acceleration*, 10.

35. A. Malm and A. Hornborg, "The Geology of Mankind? A Critique of the Anthropocene Narrative," *Anthropocene Review* 1, no. 1 (2014): 62–69; see also A. Malm, *Fossil Capital: The Rise of the Steam Engine and the Roots of Global Warming* (London: Verso, 2016).

36. Malm, *Fossil Capital*, 31.

37. Malm and Hornborg, "Geology of Mankind?," 65.

38. A. Acker, "A Different Story of the Anthropocene: Brazil's Post-colonial Quest for Oil (1930–1975)," *Past and Present* 249, no. 1 (2020): 172; E. Chatterjee, "The Asian Anthropocene: Electricity and Fossil Developmentalism," *Journal of Asian Studies* 79,

no. 1 (2020): 3–24; V. Seow, *Carbon Technocracy: Energy Regimes in Modern East Asia* (Chicago: University of Chicago Press, 2022).

39. P. S. Sutter, "Putting the Intellectual Back in Environmental History," *Modern Intellectual History* 18, no. 2 (2020): 596.

40. Sutter, "Putting the Intellectual Back." Sutter acknowledges two recent publications that have directly engaged with intellectual history: P. Warde, L. Robin, and S. Sörlin, *The Environment: A History of the Idea* (Baltimore: Johns Hopkins University Press, 2019); and K. M. Woodhouse, *The Ecocentrists: A History of Radical Environmentalism* (New York: Columbia University Press, 2020).

41. J. E. Toews, "Intellectual History after the Linguistic Turn: The Autonomy of Meaning and the Irreducibility of Experience," *American History Review* 92, no. 4 (1987): 879–907.

42. Even though work within the energy humanities has engaged directly with intellectual history, to date the focus has remained largely on individual thinkers such as George Bataille. See A. Stoekl, *Bataille's Peak: Energy, Religion, and Postsustainability* (Minneapolis: University of Minnesota Press, 2007).

43. C. F. Jones, "Petromyopia: Oil and the Energy Humanities," *Humanities* 5, no. 2 (2016): 1–10.

44. Recent debates about what American studies could bring to energy history also raise the potential to engage further with intellectual history. American studies would provide a rich disciplinary foundation to unravel how energy has shaped American culture, institutions, the material praxis, and belief systems. For more on this debate, see M. Schneider-Mayerson, "Necrocracy in America: American Studies Begins to Address Fossil Fuels and Climate Change," *American Quarterly* 67, no. 2 (2015): 529–540; M. Ziser, N. Zaretsky, and J. Sze, "Perpetual Motion: Energy and American Studies," *American Quarterly* 72, no. 3 (2020): 543–557. Caleb Wellum has also argued for a more expansive role of energy history in reimagining some of the key narratives in American history. C. Wellum, "The Use of Energy History," *Modern American History* 6 (2023): 201–219.

45. P. Mirowski, *More Heat Than Light: Economics as Social Physics, Physics as Nature's Economics* (Cambridge: Cambridge University Press, 1989), 127–132.

46. Rabinbach, *Human Motor*, 3.

47. Rabinbach, *Human Motor*, 3–5.

48. Rabinbach, *Human Motor*, 3.

49. P. Mirowski, *The Effortless Economy of Science* (Durham, NC: Duke University Press), 154.

50. For an account of how the new engineering ethos transformed American liberal thought in the twentieth century, see J. M. Jordan, *Machine-Age Ideology: Social Engineering and American Liberalism, 1911–1939* (Chapel Hill: University of North Carolina Press, 1994).

51. Morton Keller puts this impetus at the center of progressive politics, arguing that it united figures across the ideological spectrum, upsetting traditional divisions between the Left and the Right. M. Keller, *Regulating a New Economy: Public Policy and Social Change in America, 1900–1933* (Cambridge, MA: Harvard University Press, 1994), 4.

52. For the cultural expression of this "machine age consciousness," see C. Tichi, *Shifting Gears: Technology, Literature, Culture in Modernist America* (Chapel Hill:

University of North Carolina Press, 1987); R. G. Wilson, D. H. Pilgrim, and D. Tashjian, *The Machine Age in America, 1918–1941* (New York: Brooklyn Museum, 1986).

53. Jordan, *Machine-Age Ideology*, 2.

54. Martha Banta has described how this engineering ethos fed into the social narratives of the period. See M. Banta, *Taylored Lives: Narrative Productions in the Age of Taylor, Veblen, and Ford* (Chicago: University of Chicago Press, 1993).

55. For an account of how objectivity fed into the social sciences, see D. Ross, *The Origins of American Social Science* (Cambridge: Cambridge University Press, 1991); M. C. Smith, *Social Science in the Crucible: The American Debate over Objectivity and Purpose, 1918–1941* (Durham, NC: Duke University Press, 1994); A. Jewett, *Science, Democracy, and the American University: From the Civil War to the Cold War* (Cambridge: Cambridge University Press, 2012).

56. "Germany Has Been Prussianized, Says Bergson: Famous French Philosopher Declares Admiration of Brute Force Has Become a National Ideal," *New York Times*, January 10, 1915, 43.

57. "Germany Has Been Prussianized," 43.

58. J. Dewey, "On Understanding the Mind of Germany," *Atlantic Monthly*, February 1916, 259.

59. Dewey, "On Understanding the Mind," 259.

60. Dewey, "On Understanding the Mind," 260.

61. S. Freud, "Beyond the Pleasure Principle," in *The Standard Edition of the Complete Psychological Works of Sigmund Freud*, vol. 18, *(1920–1922): Beyond the Pleasure Principle, Group Psychology, and Other Works*, ed. J. Strachey et al. (London: Hogarth, 1955), 41.

62. For a description of how Freud drew on energy within his metapsychology, see N. Mansfield, *The God Who Deconstructs Himself: Sovereignty and Subjectivity between Freud, Bataille, and Derrida* (New York: Fordham University Press, 2010).

63. Sergio Franzese has argued for the centrality of energy within James's philosophy. See S. Franzese, *The Ethics of Energy: William James's Moral Philosophy in Focus* (Frankfurt: Ontos, 2008).

64. W. James, "The Energies of Men" (1906), in *Writings, 1902–1910*, ed. B. Kuklick (New York: Literary Classics of the United States, 1987).

65. James, "Energies of Men," 1239.

66. James, "Energies of Men," 1239–1240.

67. James expanded on this point in a revised version of the talk published in the *Philosophical Review* in 1907. W. James, *The Energies of Men* (New York: Moffat, Yard, 1914), 34.

68. W. James, "Pragmatism's Conception of Truth" (1907), in *Pragmatism, and Four Essays from "The Meaning of Truth"* (Cleveland: Meridian Books, 1942), 141–142.

69. W. James, "The Moral Equivalent of War" (1910), in W. James, *Writings, 1902–1910* (New York: Literary Classics of the United States, 1987), 1281–1293.

70. W. James, *The Varieties of Religious Experience* (Cambridge, MA: Harvard University Press, 1985), 292.

71. James, "Moral Equivalent of War," 1281–1293.

72. W. Lippmann, *A Preface to Politics* (New York: Mitchell Kennerley, 1913), 50.

73. Lippmann, *Preface to Politics*, 50.

74. W. James to H. Adams, June 26, 1910, in *The Letters of William James*, ed. H. James (Boston: Atlantic Monthly Press, 1920), 2:347.

75. For more on how energy appealed to the monist worldview, see T. H. Weir, "The Riddles of Monism: An Introductory Essay," in *Monism Science, Philosophy, Religion, and the History of a Worldview*, ed. T. H. Weir (New York: Palgrave Macmillan, 2012), 6–7.

76. W. Ostwald, *Monism as the Goal of Civilization* (Hamburg: International Committee of Monism, 1913), 30.

77. See J. K. Alexander, *The Mantra of Efficiency: From Waterwheel to Social Control* (Baltimore: Johns Hopkins University Press, 2010), 2.

78. See D. Rodgers, *Atlantic Crossings: Social Politics in a Progressive Age* (Cambridge, MA: Harvard University Press, 1998).

79. E. K. Spann, *Designing Modern America: The Regional Planning Association of America and Its Members* (Columbus: Ohio State University Press, 1996), 80.

Chapter 1 · *Energizing the Will*

1. J. Loeb to T. Roosevelt, February 8, 1909, box 13, Jacques Loeb Papers, Manuscript Division, Library of Congress, Washington, DC (hereafter cited as Loeb Papers).

2. See H. Fangerau, "From Mephistopheles to Isaiah: Jacques Loeb, Technical Biology and War," *Social Studies of Science* 39, no. 2 (2009): 229–256.

3. Loeb was active in protesting America's involvement in the First World War. He was a member of four different antiwar organizations: the American League to Limit Armaments, the American Neutral Conference Committee, the Emergency Peace Federation, and the People's Council of America for Democracy and Peace. C. Rasmussen and R. Tilman, *Jacques Loeb: His Science and Social Activism and Their Philosophical Foundations* (Philadelphia: American Philosophical Society, 1998), 72–73.

4. Loeb to Roosevelt, February 8, 1909.

5. Loeb to Roosevelt, February 8, 1909.

6. Loeb to Roosevelt, February 8, 1909.

7. The connection between Roosevelt's conservation agenda and the "efficiency movement" has been well documented. See S. Hays, *Conservation and the Gospel of Efficiency: The Progressive Conservation Movement, 1890–1920* (Pittsburgh: University of Pittsburgh Press, 1999); I. E. Tyrrell, *Crisis of the Wasteful Nation: Empire and Conservation in Theodore Roosevelt's America* (Chicago: University of Chicago Press, 2015). See also B. H. Johnson, *Escaping the Dark, Gray City: Fear and Hope in Progressive-Era Conservation* (New Haven, CT: Yale University Press, 2017).

8. T. Roosevelt to J. Loeb, February 13, 1909, box 13, Loeb Papers.

9. See T. Roosevelt, "The Strenuous Life: Speech before the Hamilton Club, Chicago, April 10, 1899" (1899), in *The Strenuous Life: Essays and Addresses* (Mineola, NY: Dover, 2009), 1–10.

10. There is an extensive literature on Roosevelt's promotion of the strenuous life. See L. G. Dorsey, *We Are All Americans, Pure and Simple: Theodore Roosevelt and the Myth of Americanism* (Tuscaloosa: University of Alabama Press, 2007); S. L. Watts, *Rough Rider in the White House: Theodore Roosevelt and the Politics of Desire* (Chicago: University of Chicago Press, 2003); K. Dalton, "Why America Loved Teddy Roosevelt: Or, Charisma Is in the Eyes of the Beholders," in *Our Selves/Our Past: Psychological*

Approaches to American History, ed. R. Brugger (Baltimore: Johns Hopkins University Press, 1981), 269–291.

11. Elihu Root, quoted in K. Dalton, *Theodore Roosevelt: A Strenuous Life* (New York: Alfred A. Knopf, 2002), 202; J. Burroughs, *Camping and Tramping with President Roosevelt* (Carlisle, MA: Applewood Books, 1907), 60.

12. J. Loeb to A. Pinchot, February 17, 1917, box 11, Loeb Papers. In 1915 Loeb would express his fears to the chemist Wilhelm Ostwald about the increasing militarization of the United States. J. Loeb to W. Ostwald, July 31, 1915, box 11, Loeb Papers.

13. J. Loeb, "The Price of Romanticism" (n.d.), box 44, Loeb Papers.

14. B. Johnson, *Carbon Nation: Fossil Fuels in the Making of American Culture* (Lawrence: University Press of Kansas, 2014), 15.

15. Johnson, *Carbon Nation*, 5.

16. See C. T. de la Peña, *The Body Electric: How Strange Machines Built the Modern American* (New York: New York University Press, 2003); T. Lutz, *American Nervousness, 1903: An Anecdotal History* (Ithaca, NY: Cornell University Press, 1991); S. Kern, *The Culture of Time and Space, 1880–1918* (Cambridge, MA: Harvard University Press, 2003); D. Nye, *Electrifying America: Social Meanings of a New Technology, 1880–1940* (Cambridge, MA: MIT Press, 1992); D. Nye, *Consuming Power: A Social History of American Energies* (Cambridge, MA: MIT Press, 1998).

17. This talk was published as the introductory essay of the book *The Mechanistic Conception of Life*. J. Loeb, "The Mechanistic Conception of Life" (1912), in *The Mechanistic Conception of Life* (Cambridge, MA: Belknap Press of Harvard University Press, 1964), 30.

18. Loeb, "Mechanistic Conception of Life," 30.

19. See P. J. Pauly, *Controlling Life: Jacques Loeb and the Engineering Ideal in Biology* (New York: Oxford University Press, 1987), 34–37.

20. Pauly, *Controlling Life*, 37. For more on Gustav Fechner, see M. Heidelberger, *Nature from Within: Gustav Theodor Fechner and His Psychophysical Worldview* (Pittsburgh: University of Pittsburgh Press, 2004).

21. Pauly, *Controlling Life*, 37.

22. Pauly, *Controlling Life*, 37–39.

23. J. Loeb, "The Significance of Tropisms for Psychology" (1909), in *The Mechanistic Conception of Life* (Cambridge, MA: Belknap Press of Harvard University Press, 1964), 41.

24. Loeb discussed Hammond's inventions in J. Loeb, "Mechanistic Science and Metaphysical Romance," *Yale Review* 4 (July 1915): 781.

25. Loeb, "Mechanistic Conception of Life," 30.

26. Loeb, "Mechanistic Conception of Life," 30.

27. Loeb, "Mechanistic Conception of Life," 30.

28. Loeb, "Mechanistic Conception of Life," 31.

29. "The Mechanistic Conception of Life," *Daily People*, November 19, 1912; "The Mechanistic Conception of Life," *Inquirer*, December 28, 1912; "The Mechanistic Conception of Life," *Athenaeum*, October 19, 1912; "The Mechanistic Conception of Life," *San Francisco Chronicle*, September 15, 1912; "The Mechanistic Conception of Life," *Scotsman*, October 28, 1912; "The Mechanistic Conception of Life," *London Times*, September 19, 1912; "The Mechanistic Conception of Life," *Boston Transcript*,

November 8, 1912; "The Mechanistic Conception of Life," *Nation*, December 5, 1912; "The New Realism—a Philosophy of Disillusionment," *Current Opinion*, January 1913, 52; "The Mechanistic Conception of Life," *Chicago Evening Post*, February 28, 1913; "The Mechanistic Conception of Life," *Hebrew Standard*, July 11, 1913.

30. C. Teller, "Man as Mechanism: The Extraordinary Experiments of Jacques Loeb," *Metropolitan Magazine*, November 1912.

31. A. S. Baird, "Wonderful Findings of College Professors," *Nashville Banner, Tennessee*, January 18, 1913, 11.

32. "Morality More Than an Instinct: Dr. Neumann Attacks Materialism," *Brooklyn Citizen*, December 9, 1912.

33. C. Snyder, "Theory of Life: Dr. Jacques Loeb's Mechanistic Conception of Energy," *New York Times*, October 6, 1912, BR554.

34. Snyder, "Theory of Life," BR554.

35. F. Burwick and P. Douglass, introduction to *The Crisis in Modernism: Bergson and the Vitalist Controversy*, ed. F. Burwick and P. Douglass (Cambridge: Cambridge University Press, 1992), 3.

36. R. Pearl, "The Living Machine," *Dial* 54, no. 638 (January 16, 1913): 51.

37. Pearl, "Living Machine," 51.

38. Pearl, "Living Machine," 51.

39. For more on Loeb's relationship to the German Monist League, see H. Fangerau, "Monism, Racial Hygiene, and National Socialism," in *Monism, Science, Philosophy, Religion, and the History of a Worldview*, ed. T. H. Weir (New York: Palgrave Macmillan, 2012), 223–249.

40. It is important to note that the Monist League was strongly influenced by Haeckel's social Darwinism, embodying racial theories that would later inform the ideology of National Socialism. See D. Gasman, *The Scientific Origins of National Socialism: Social Darwinism in Ernst Haeckel and the German Monist League* (London: Macdonald, 1971).

41. Gasman, *Scientific Origins of National Socialism*, 89.

42. W. Ostwald, *Monism as the Goal of Civilization* (Hamburg: International Committee of Monism, 1913).

43. Ostwald, *Monism as the Goal*, 31.

44. Ostwald, *Monism as the Goal*, 30.

45. For more on Ostwald's cultural activities, see N. R. Holt, "Wilhelm Ostwald's 'The Bridge,'" *British Journal for the History of Science* 10, no. 2 (1977): 146–150; T. Hapke, "Wilhelm Ostwald, the Brücke (Bridge), and Connections to Other Bibliographic Activities at the Beginning of the Twentieth Century," in *Proceedings of the 1998 Conference on the History and Heritage of Science Information Systems*, ed. M. E. Bowden, T. B. Hahn, and R. V. Williams (Medford, NJ: Information Today, 1999), 139–147; J. Stewart, "Sociology, Culture and Energy: The Case of Wilhelm Ostwald's 'Sociological Energetics'—a Translation and Exposition of a Classic Text," *Cultural Sociology* 8, no. 3 (2014): 333–350.

46. "Herr Professor Ostwald Wants to Organize a Brain Trust," *Milwaukee Journal*, March 9, 1913, 8.

47. Ostwald, *Monism as the Goal*, 33; T. H. Weir, "The Riddles of Monism: An Introductory Essay," in *Monism Science, Philosophy, Religion, and the History of a Worldview*, ed. T. H. Weir (New York: Palgrave Macmillan, 2012), 7.

48. Ostwald, Monism as the Goal, 31, 37.
49. Ostwald, Monism as the Goal, 37.
50. For more about this debate, see R. Deltete, "Helm and Boltzmann: Energetics at the Lübeck Naturforscherversammlung," Synthese 119, no. 1/2 (1999): 45–68.
51. "Energy of the Human Will," Literary Digest 36, no. 360 (February 15, 1908): 223.
52. "Energy of the Human Will," 223.
53. W. E. A. Wilkinson, "Will-Force and the Conservation of Energy," Monist 18, no. 1 (1908): 1.
54. Wilkinson, "Will-Force," 9.
55. W. P. Montague, "Are Mental Processes in Space?," Monist 18, no. 1 (1908): 21–29.
56. P. Carus, "A Monistic Conception of Consciousness: In Reply to Mr. Ayton Wilkinson's Article on 'Will-Force' and Mr. Montague's 'Are Mental Processes in Space?,'" Monist 18, no. 1 (1908): 35, 39.
57. C. Smith and M. N. Wise, Energy and Empire: A Biographical Study of Lord Kelvin (Cambridge: Cambridge University Press, 1989), 617. See also L. Daston, "The Theory of Will versus the Science of Mind," in The Problematic Science: Psychology in Nineteenth-Century Thought, ed. W. R. Woodward and M. G. Ash (New York: Praeger, 1982), 107–108.
58. J. Tyndall, Address Delivered before the British Association Assembled at Belfast, with Additions (London: Longmans, Green, 1874), 45.
59. For more on this debate, see Smith and Wise, Energy and Empire, 612–632.
60. Smith and Wise, Energy and Empire, 625.
61. J. C. Maxwell, Theory of Heat (London: Longmans, Green, 1871), 308–309, quoted in Smith and Wise, Energy and Empire, 625.
62. Smith and Wise, Energy and Empire, 627.
63. R. Smith, Free Will and the Human Sciences in Britain, 1870–1910 (London: Pickering & Chatto, 2013).
64. G. Holton, "Ernst Mach and the Fortunes of Positivism in America," Isis 83, no. 1 (1992): 29.
65. E. Mach, History and Root of the Principle of the Conservation of Energy, trans. Philip E. B. Jourdain (1886; repr., Chicago: Open Court, 1911); E. Mach, Principles of the Theory of Heat: Historically and Critically Elucidated, ed. B. McGuinness (Dordrecht: D. Reidel, 1986).
66. See Mach, Principles of the Theory.
67. For more on the relationship between Mach and Ostwald, see M. Neuber, "Uneasy Allies: Ostwald, Helm, Mach and Their Philosophies of Science," in Wilhelm Ostwald at the Crossroads between Chemistry, Philosophy and Media Culture, ed. B. Görs, N. Psarros, and P. Ziche (Leipzig: Leipziger Universitätsverlag, 2005), 47–58.
68. W. Ostwald, "The Philosophical Meaning of Energy," International Quarterly 7 (March 1903): 300–315.
69. Ostwald's talk was reported in T. L. Smith, "Notes and News," American Journal of Psychology 17, no. 1 (January 1906): 146.
70. Smith, "Notes and News," 146.
71. For more on the relationship between Ostwald's energetics, monism, and ethics, see D. Sobczynska and E. Czerwinska, "Wilhelm Ostwald and the German Monistic League on the Social and Cultural Role of Science," in Wilhelm Ostwald at the Crossroads

between Chemistry, Philosophy and Media Culture, ed. B. Görs, N. Psarros, and P. Ziche (Leipzig: Leipziger Universitätsverlag, 2005), 59–68.

72. The popular American press reported extensively on Ostwald's energetics, especially during his stay at Harvard University in 1906. For example, the popular science writer E. E. Slosson, in his book *Major Prophets of Today* (1914), held Ostwald up as one of the most important thinkers of his generation. E. E. Slosson, *Major Prophets of Today* (Boston: Little, Brown, 1914), 190–242.

73. W. Ostwald, *Individuality and Immortality* (Boston: Houghton, Mifflin, 1906), 42–43.

74. Ostwald, *Individuality and Immortality*, 43.

75. Ostwald, *Individuality and Immortality*, 44.

76. Ostwald, *Individuality and Immortality*, 44.

77. Ostwald, *Individuality and Immortality*, 45–46. Elsewhere, Ostwald developed a formula for describing happiness, whereby the degree of happiness was dependent on an inverse relation between voluntary and involuntary resistances asserted on the will. W. Ostwald, "A Theory of Happiness," *International Quarterly* 11 (April 1, 1905): 316–326.

78. Ostwald, *Individuality and Immortality*, 46.

79. Ostwald, *Individuality and Immortality*, 71.

80. Pauly, *Controlling Life*, 42.

81. H. Fangerau, "Biology and War—American Biology and International Science," *History and Philosophy of the Life Sciences* 29, no. 4 (2007): 404.

82. Fangerau, "Biology and War," 405.

83. Loeb, "Mechanistic Science and Metaphysical Romance," 785.

84. Loeb, "Price of Romanticism" (n.d.).

85. Loeb, "Mechanistic Science and Metaphysical Romance," 785.

86. C. T. Rasmussen and R. Tilman, "Mechanistic Physiology and Institutional Economics: Jacques Loeb and Thorstein Veblen," *International Journal of Social Economics* 19, no. 10/11/12 (1992): 235–247.

87. Rasmussen and Tilman, *Jacques Loeb*, 93.

88. T. Veblen, *The Theory of the Leisure Class* (Boston: Houghton Mifflin, 1973), 40.

89. Veblen, *Theory of the Leisure Class*, 31.

90. Veblen, *Theory of the Leisure Class*, 152.

91. T. Veblen, *The Instinct of Workmanship, and the State of the Industrial Arts* (New York: Macmillan, 1914), 334n1. Rasmussen and Tilman point out how Veblen shared Loeb's disdain of Bergson's thought. Rasmussen and Tilman, *Jacques Loeb*, 119.

92. Veblen, *Instinct of Workmanship*, 334.

93. Veblen, *Instinct of Workmanship*, 336–340.

94. See M. G. Land, "Three Max Gottliebs: Lewis's, Dreiser's, and Walker Percy's View of the Mechanist-Vitalist Controversy," *Studies in the Novel* 15, no. 4 (1983): 314–331.

95. G. W. Cooke, "Science and Socialism," *Wall Street Journal*, September 25, 1912.

96. See G. Griffith, *Socialism and Superior Brains: The Political Thought of Bernard Shaw* (London: Routledge, 1993), 125–131.

97. P. J. Hale, "The Search for Purpose in a Post-Darwinian Universe: George Bernard Shaw, 'Creative Evolution,' and Shavian Eugenics: 'The Dark Side of the Force,'" *History and Philosophy of the Life Sciences* 28, no. 2 (2006): 191–213.

98. H. F. Osborn, *The Origin and Evolution of Life: On the Theory of Action, Reaction and Interaction of Energy* (New York: Charles Scribner's Sons, 1917), 17.

99. Osborn, *Origin and Evolution of Life*, 17.

100. Osborn, *Origin and Evolution of Life*, 20.

101. T. H. Morgan to H. F. Osborn, December 26, 1917, quoted in G. E. Allen, "T. H. Morgan and the Emergence of a New American Biology," *Quarterly Review of Biology* 44, no. 2 (1969): 179–180.

102. B. Regal, *Henry Fairfield Osborn: Race, and the Search for the Origins of Man* (Aldershot: Ashgate, 2002), 64.

103. Osborn, *Origin and Evolution of Life*, 10.

104. Osborn, *Origin and Evolution of Life*, 10.

105. "The Most Baffling Fact of Human Knowledge: We May Have to Give Up Matter and Form Conceptions or Be Unscientific," *Current Opinion* 65, no. 1 (July 1918): 35.

106. "Most Baffling Fact," 35.

107. J. Loeb to T. H. Morgan, February 19, 1918, Loeb Papers, box 9.

108. Morgan to Osborn, December 26, 1917.

109. Morgan to Osborn, December 26, 1917.

110. Morgan to Osborn, December 26, 1917.

111. Morgan to Osborn, December 26, 1917.

112. Morgan to Osborn, December 26, 1917.

113. Morgan to Osborn, December 26, 1917.

114. T. Roosevelt, "The Origin and Evolution of Life," *Outlook*, January 16, 1918, 97.

115. Roosevelt, "Origin and Evolution of Life," 97.

116. Roosevelt, "Origin and Evolution of Life," 97.

117. As an example, see A. W. Smith to H. F. Osborn, July 5, 1918, and W. Brehmer to H. F. Osborn, January 14, 1924, box 105, folder 1, Henry Fairfield Osborn Papers, American Museum of Natural History.

118. Regal, *Henry Fairfield Osborn*, 111.

119. For more on how Spencer popularized a view of social progress driven by force, see R. E. Martin, *American Literature and the Universe of Force* (Durham, NC: Duke University Press, 1981). For more on the connection between the second law and fin de siècle theories of social degeneration, see M. Tattersall, "Thermal Degeneration: Thermodynamics and the Heat-Death of the Universe in Victorian Science, Philosophy, and Culture," in *Decadence, Degeneration and the End: Studies in the European Fin de Siècle*, ed. M. Härmänmaa and C. Nissen (New York: Palgrave Macmillan, 2014), 17–34.

120. G. Beard, *American Nervousness: Its Causes and Consequences: A Supplement to Nervous Exhaustion (Neurasthenia)* (New York: G. P. Putnam's Sons, 1881). There is an extensive literature on neurasthenia, both in the United States and in Europe. See F. G. Gosling, *Before Freud: Neurasthenia and the American Medical Community, 1870–1910* (Champaign: University of Illinois Press, 1987); D. G. Schuster, *Neurasthenic Nation: America's Search for Health, Happiness, and Comfort, 1869–1920* (New Brunswick, NJ: Rutgers University Press, 2011); Lutz, *American Nervousness*; A. Rabinbach, *The Human Motor: Energy, Fatigue, and the Origins of Modernity* (New York: Basic Books, 1990); A. Killen, *Berlin Electropolis: Shock, Nerves, and German Modernity* (Berkeley: University of California Press, 2006); de la Peña, *Body Electric*; M. J. Cowan, *Cult of the Will:*

Nervousness and German Modernity (University Park: Pennsylvania State University Press, 2008).

121. Beard, *American Nervousness*, 9–13.

122. H. Adams to C. M. Gaskell, August 2, 1910, in *Letters of Henry Adams*, ed. W. C. Ford (Boston: Houghton Mifflin, 1938), 2:546. Quoted in G. Cotkin, *William James, Public Philosopher* (Champaign: University of Illinois Press, 1994), 75.

123. Cowan, *Cult of the Will*, 9. Tom Lutz, for example, has described how neurasthenia mirrored contemporary economic discourse. See Lutz, *American Nervousness*.

124. Beard, *American Nervousness*, 9–13.

125. For more on these therapeutic methods, see Cowan, *Cult of the Will*.

126. B. Adams, *The Law of Civilization and Decay: An Essay on History* (1895; repr., London: Macmillan, 1910), viii.

127. Adams, *Law of Civilization and Decay*, ix.

128. Adams, *Law of Civilization and Decay*, ix.

129. Adams, *Law of Civilization and Decay*, x.

130. Adams, *Law of Civilization and Decay*, ix.

131. Adams, *Law of Civilization and Decay*, 383.

132. H. Adams, *A Letter to American Teachers of History* (Washington, DC: J. H. Furst, 1910), 98–106.

133. Adams, *Letter to American Teachers*, 101.

134. Adams, *Letter to American Teachers*, 145.

135. Adams, *Letter to American Teachers*, 111–112. Although I am stressing the difference between the brothers here, both Henry Adams and Brooks Adams remained ambivalent about the process of modernity and put forward different models for understanding it. See T. J. Jackson Lears, *No Place of Grace: Antimodernism and the Transformation of American Culture, 1880–1920* (New York: Pantheon Books, 1981). See also K. N. Hayles's discussion of Henry Adams in *Chaos Bound: Orderly Disorder in Contemporary Literature and Science* (Ithaca, NY: Cornell University Press, 1990).

136. For more on the different ways that Henry Adams and Brooks Adams applied the energy laws, see T. M. Donovan, *Henry Adams and Brooks Adams: The Education of Two American Historians* (Norman: University of Oklahoma Press, 1961), 150–157.

137. T. Roosevelt, "The Law of Civilization and Decay," *Forum* 22 (January 1897): 588–589.

138. Brooks Adams's role in American foreign policy has been well documented. See W. LaFeber, *The New Empire: An Interpretation of American Expansion, 1860–1898* (Ithaca, NY: Cornell University Press, 1998), 84–85; William A. Williams, "Brooks Adams and American Expansion," *New England Quarterly* 25, no. 2 (June 1952): 217–232.

139. LaFeber, *New Empire*, 84.

140. For the evolution of Brooks Adams's thought, see M. J. Carson, "The Evolution of Brooks Adams," *Biography* 6, no. 2 (1983): 95–116.

141. B. Adams, "The Heritage of Henry Adams" (1919), in H. Adams, *The Degradation of the Democratic Dogma* (New York: Harper & Row, 1969), 105.

142. Adams, "Heritage of Henry Adams," 105.

143. Adams, "Heritage of Henry Adams," 84.

144. Adams, "Heritage of Henry Adams," 109.

145. Much has been made of the fact that James suffered from neurasthenia himself and the potential impact this had on his pragmatist philosophy. See G. Cotkin, *William James, Public Philosopher* (Baltimore: Johns Hopkins University Press, 1990), 40–73.

146. W. James to H. Adams, June 17, 1910, quoted in G. Monteiro, "Henry Adams and William James: An Intellectual Relationship," *American Literary Realism* 39, no. 2 (2007): 162. See also J. P. Diggins, *The Promise of Pragmatism: Modernism and the Crisis of Knowledge and Authority* (Chicago: University of Chicago Press, 1994), 109.

147. James to Adams, June 17, 1910, 162.

148. James to Adams, June 17, 1910, 162.

149. James to Adams, June 17, 1910, 162.

150. James to Adams, June 17, 1910, 162.

151. W. James, "Herbert Spencer," *Atlantic Monthly*, July 1904, 99–108.

152. James, "Herbert Spencer," 99–108.

153. James, "Herbert Spencer," 107.

154. James, "Herbert Spencer," 107.

155. James, "Herbert Spencer," 107.

156. James to Adams, June 17, 1910, 162.

157. W. James, *The Principles of Psychology* (Cambridge, MA: Harvard University Press, 1981), 2:1132–1133.

158. W. James, *The Varieties of Religious Experience* (1902; repr., Cambridge, MA: Harvard University Press, 1985), 162. For this interpretation of how James incorporated energy into his philosophical program, I am indebted to S. Franzese, *The Ethics of Energy: William James's Moral Philosophy in Focus* (Frankfurt: Ontos, 2008).

159. James, *Varieties of Religious Experience*, 162.

160. James, *Varieties of Religious Experience*, 173.

161. W. James, "The Energies of Men" (1906), in *William James, Writings, 1902–1910*, ed. B. Kuklick (New York: Literary Classics of the United States, 1987), 1226.

162. James, "Energies of Men," 1226.

163. James, "Energies of Men," 1224.

164. James, "Energies of Men," 1227.

165. James, "Energies of Men," 1234. Tom Lutz has argued that James drew on the language of consumer capitalism to promote an ethics of "overspending," where energy was no longer tied to physical limits but could expand infinitely. See Lutz, *American Nervousness*, 75.

166. James, "Energies of Men," 1226.

167. James, "Energies of Men," 1227, 1230.

168. James, "Energies of Men," 1239.

169. For more on this connection, see K. L. Hoganson, *Fighting for American Manhood: How Gender Politics Provoked the Spanish-American and Philippine-American Wars* (New Haven, CT: Yale University Press, 1998).

170. See Lutz, *American Nervousness*; Watts, *Rough Rider*.

171. Franzese, *Ethics of Energy*, 160.

172. T. Roosevelt, "Man and Statehood" (1901), in *The Strenuous Life: Essays and Addresses* (Mineola, NY: Dover, 2009), 118.

173. Watts, *Rough Rider*, 19.

174. For more on the contrast between Roosevelt's and James's understanding of the strenuous life, see P. K. Dooley, "Public Policy and Philosophical Critique: The William James and Theodore Roosevelt Dialogue on Strenuousness," *Transactions of the Charles S. Peirce Society* 37, no. 2 (2001): 161–177.

175. See Roosevelt, "Strenuous Life."

176. W. James, letter to the editor, April 15, 1899, *Springfield Republican*, quoted in Dooley, "Public Policy and Philosophical Critique," 171.

177. According to Sergio Franzese, James refused to take energy as a monolithic principle. Instead, he maintained that energy only had ethical value as something that enabled action and was organized and managed in the act. Franzese, *Ethics of Energy*, 198.

178. Franzese, *Ethics of Energy*, 199.

179. W. J. McGee and N. C. Blanchard, eds., *Proceedings of a Conference of Governors in the White House, Washington, D.C., May 13–15, 1908* (Washington, DC: US Government Printing Office, 1909).

180. See A. Carnegie, "The Conservation of Ores and Related Minerals," in McGee and Blanchard, *Proceedings of a Conference*, 14–25; S. Gompers, "Conservation in Relation to Labor," in McGee and Blanchard, *Proceedings of a Conference*, 398–403; R. O. Richards, "Methods for Conservation," in McGee and Blanchard, *Proceedings of a Conference*, 373–376.

181. Watts, *Rough Rider*, 52.

182. For a discussion of the influence of Stanley Hall and Edward Ross on Roosevelt, see Watts, *Rough Rider*, 51–57.

183. E. A. Ross, *Social Control: A Survey of the Foundations of Order* (London: Macmillan, 1901), 365.

184. Ross, *Social Control*, 365.

185. W. Ostwald, quoted in A. L. F., "A Letter to a German-American Friend," *Outlook*, January 6, 1915, 16.

186. See S. Haber, *Efficiency and Uplift: Scientific Management in the Progressive Era, 1890–1920* (Chicago: University of Chicago Press, 1964), 119–120; G. R. Searle, *The Quest for National Efficiency: A Study in British Politics and Political Thought, 1899–1914* (Berkeley: University of California Press, 1971), 54–57.

187. A. L. F., "Letter to a German-American Friend," 13–16. See also H. G. Wells, "Ideals of Organization," *New Republic* 3, no. 38 (July 24, 1915): 301–302; "Democracy in the Fining-Pot," *Literary Digest* 51, no. 7 (August 15, 1915): 298–299.

188. "Professor Ostwald and the University of Leipsic," *Outlook*, February 3, 1915, 284.

189. B. Adams, *The Theory of Social Revolutions* (New York: Macmillan, 1913), 5.

190. T. Anderson, *Brooks Adams, Constructive Conservative* (Ithaca, NY: Cornell University Press, 1951), 167.

191. Adams, *Theory of Social Revolutions*, 6.

192. Adams, *Theory of Social Revolutions*, 6.

193. W. Wilson, *The New Freedom: A Call for the Emancipation of the Generous Energies of a People* (New York: Doubleday, Page, 1913), 292.

194. Wilson, *New Freedom*, 88.

195. Adams, "Heritage of Henry Adams," 115–116.

196. Anderson, *Brooks Adams*, 170.

197. Anderson, *Brooks Adams*, 170.

198. Brooks Adams to Henry Cabot Lodge, January 18, 1920, quoted in Anderson, *Brooks Adams*, 180.

Chapter 2 · *Energizing Americanness*

1. Sections of this chapter were published as R. Wright, "Sunspots and Sync," in *Energy in Literature: Essays on Energy and Its Social and Environmental Implications in Twentieth and Twenty-First Century Literary Texts*, ed. P. A. Farca (Oxford: TrueHeart, 2015). See A. L. Tchijevsky, "Physical Factors of the Historical Process" (1926), trans. V. P. De Smitt, *Cycles*, January 1971, 13. Tchijevsky's paper was first presented to the American Meteorological Society, which was meeting as part of the American Association for the Advancement of Science, on December 31, 1926. It was read and translated by Vladimir P. De Smitt, a research associate at Columbia University. Tchijevksy was the anglicized version of his Russian name, Alexander Leonidovich Chizhevsky.

2. Tchijevsky, "Physical Factors," 14.

3. "Man Is Slave of Sun-Spots Scientist Says," *Courier-Post*, December 31, 1926, 1; "Man Slave of Sun, Avers Scientist," *Brooklyn Daily Times*, December 31, 1926, 1; "Scientist Declares World Is Held in Slavery by Sun," *Evening Sun* (Hannover, Pennsylvania), December 31, 1926, 8; "Mass Excitement Laid to Sunspots; Dr V. P. De Smitt Says Russian Scientist Has Traced World Unrest to Solar Phenomena," *New York Times*, March 8, 1927, 52; "Sunspots Called Key to the Future; Russian Professor Finds Parallel between Human and Solar Activities and Would Time Events Accordingly," *New York Times*, January 30, 1927, xx9; "Sunspots Cause of Wars?," *Popular Science* 110, no. 6 (June 1927): 50.

4. See Tchijevsky, "Physical Factors," 21.

5. In the *New York Times* article, the journalist reported Tchijevsky's talk at the American Meteorology Society somewhat disparagingly. He used scare quotes to discuss Tchijevsky's "revelations," highlighting that "just what form this activity is to take Tchijevsky has not attempted to prophesy!" "Sunspots Called Key," 9.

6. See, e.g., H. T. Stetson, *Sunspots and Their Effects* (New York: McGraw-Hill, 1937); C. Fitzhugh Talman, "Science Studies Effects of Sunspots on Earth: Their Influence on Ultra-violet Radiation," *New York Times*, July 17, 1927, xx4; "Sunspots Grow, and Studies Suggest Effect Human Conduct," *Evening News, Tonawanda*, February 24, 1938, 7; "Sun-Spots as Prosperity Regulator," *Literary Digest* 122, no. 4 (July 25, 1936): 16–17.

7. J. Hingham, *Strangers in the Land: Patterns of American Nativism, 1860–1925* (New Brunswick, NJ: Rutgers University Press, 2002), 264–299.

8. J. R. Fleming, *Historical Perspectives on Climate Change* (Oxford: Oxford University Press, 1998); J. R. Fleming, "Civilisation, Climate, and Ozone: Ellsworth Huntington's 'Big' Views on Biophysics, Biocosmics, and Biocracy," in *Weather, Climate and the Geographical Imagination*, ed. M. Mahony and S. Randalls (Pittsburgh: University of Pittsburgh Press, 2020), 215–231; G. J. Martin, *Ellsworth Huntington: His Life and Thought* (Hamden, CT: Archon Books, 1973); D. N. Livingstone, "Darwinian Hippocratics, Eugenic Enticements, and the Biometeorological Body," in *Weather, Climate and the Geographical Imagination*, ed. M. Mahony and S. Randalls (Pittsburgh: University of Pittsburgh Press, 2020), 191–214.

9. Fleming, *Historical Perspectives on Climate Change*, 100–105. Thomas Turnbull has also identified how the science of energy influenced geographical thought in the

twentieth century, including the work of Ellsworth Huntington. See T. Turnbull, "From Incommensurability to Ubiquity: An Energy History of Geographic Thought," *Journal of Historical Geography* 73 (2021): 11–23.

10. See C. Garcia-Mata and F. Shaffner, "Solar and Economic Relationships: A Preliminary Report," *Quarterly Journal of Economics* 49, no. 1 (1934): 1–51.

11. H. T. Stetson, "The Influence of the Sun on Human Affairs," *Scientific Monthly* 43, no. 1 (1936): 15. See also "He Knows the Sun," *Science News-Letter* 10, no. 286 (1926): 1; H. F. Newall, "George Ellery Hale. 1868–1938," *Obituary Notices of Fellows of the Royal Society* 2, no. 7 (1939): 523–529.

12. Stetson, "Influence of the Sun," 15.

13. See G. E. Webb, *Tree Rings and Telescopes: The Scientific Career of A. E. Douglass* (Tucson: University of Arizona Press, 1983).

14. S. E. Nash, "Time for Collaboration: A. E. Douglass, Archaeologists, and the Establishment of Tree-Ring Dating in the American Southwest," *Journal of the Southwest* 40, no. 3 (1998): 263.

15. A. E. Douglass, "Evidence of Climatic Effects in the Annual Rings of Trees," *Ecology* 1, no. 1 (January 1920): 27.

16. P. Anker, *Imperial Ecology: Environmental Order in the British Empire, 1895–1945* (Cambridge, MA: Harvard University Press, 2001), 98. See also C. S. Elton, "Periodic Fluctuations in the Numbers of Animals: Their Causes and Effects," *Journal of Experimental Biology* 2, no. 1 (1924): 119–163.

17. See Anker, *Imperial Ecology*, 107–108; E. Huntington, "The Matamek Conference on Biological Cycles," *Science* 74, no. 1914 (September 4, 1931): 230.

18. Huntington, "Matamek Conference on Biological Cycles," 232.

19. See E. Huntington, "The Ebb and Flow of Human Populations," in *Matamek Conference on Biological Cycles, Full Proceedings*, together with a report by Ellsworth Huntington (Canadian Labrador: Matamek Factory, 1932), 60.

20. J. Stokley, "Radio Reception and Sun Spots," *Science News-Letter* 16, no. 434 (August 3, 1929): 59.

21. Stokley, "Radio Reception and Sun Spots," 59.

22. Stetson wrote extensively about this connection. See H. T. Stetson, *Man and the Stars* (New York: Whittlesey House, McGraw-Hill, 1930); H. T. Stetson, *Earth, Radio and the Stars* (New York: Whittlesey House, McGraw-Hill, 1934); H. T. Stetson, *Sunspots and Their Effects* (New York: Whittlesey House, McGraw-Hill, 1937).

23. Stetson, *Earth, Radio and the Stars*, 7.

24. Stetson, *Earth, Radio and the Stars*, 6.

25. "Sunspots and Wild Creatures," *Literary Digest* 112, no. 11 (March 12, 1932): 32.

26. "Sunspots and Wild Creatures," 32.

27. "Sunspots and Wild Creatures," 32.

28. E. Martin, "Editor's Easy Chair: Sun Spots and Politics," *Harper's Monthly Magazine* 155, no. 923 (March 1927): 529–532; E. Martin, "Editor's Easy Chair: Sun Spots and Justice," *Harper's Monthly Magazine* 155, no. 930 (November 1927): 785–788.

29. Martin, "Sun Spots and Politics," 529.

30. Martin, "Sun Spots and Politics," 530.

31. Martin, "Sun Spots and Politics," 530.

32. Martin, "Sun Spots and Politics," 530.

33. Martin, "Sun Spots and Politics," 530.
34. Tchijevsky, "Physical Factors."
35. Tchijevsky, "Physical Factors," 21.
36. Tchijevsky, "Physical Factors," 17.
37. Tchijevsky, "Physical Factors," 16.
38. Tchijevsky, "Physical Factors," 17.
39. Tchijevsky, "Physical Factors," 18.
40. Tchijevsky, "Physical Factors," 18.
41. Tchijevsky, "Physical Factors," 21.
42. Martin, "Sun Spots and Politics," 531.
43. Martin, "Sun Spots and Politics," 531.
44. Martin, "Sun Spots and Politics," 531.
45. Martin, "Sun Spots and Justice," 785.
46. Martin, "Sun Spots and Justice," 785.
47. "The Weather and Our Feelings," *Forum* 74, no. 3 (September 1925): 390–392.
48. "Weather and Our Feelings," 391–392.
49. The physician William F. Petersen also examined the physiological effects of sunspots. See W. F. Petersen, *Lincoln, Douglas: The Weather as Destiny* (Springfield, IL: C. C. Thomas, 1943); W. F. Petersen, *Man, Weather, Sun* (Springfield, IL: C. C. Thomas, 1947).
50. D. Brett and K. M. Wertheimer, *Max Wertheimer and Gestalt Theory* (New Brunswick, NJ: Transaction, 2005), 249.
51. R. H. Wheeler, *The Laws of Human Nature* (London: Nisbet, 1931), 49.
52. R. H. Wheeler and M. G. Zahorchak, *Climate, the Key to Understanding Business Cycles, with a Forecast of Trends into the 21st Century* (Linden, NJ: Tide, 1980), 204.
53. Wheeler and Zahorchak, *Climate*, 204.
54. R. H. Wheeler, "The Effect of Climate on Human Behavior in History," *Transactions of the Kansas Academy of Science* 46 (1943): 39.
55. Wheeler, "Effect of Climate," 37.
56. Wheeler, "Effect of Climate," 45.
57. Wheeler, "Effect of Climate," 46.
58. Wheeler, "Effect of Climate," 46.
59. Wheeler, "Effect of Climate," 47.
60. Tchijevsky, "Physical Factors," 21.
61. Tchijevsky, "Physical Factors," 21.
62. Tchijevsky, "Physical Factors," 21.
63. See P. A. Sorokin, *Contemporary Sociological Theories* (New York: Harper & Brothers, 1928), 99–195.
64. P. A. Sorokin, "A Neglected Factor of War," *American Sociological Review* 3, no. 4 (August 1938): 475.
65. Sorokin, "Neglected Factor of War," 478.
66. Sorokin, *Contemporary Sociological Theories*, 36.
67. P. A. Sorokin, "Is Accurate Social Planning Possible?," *American Sociological Review* 1, no. 1 (1936): 13.
68. See Fleming, *Historical Perspectives on Climate Change*.

69. Huntington's climatic determinism was mostly rejected by the geographical community during the interwar period. Outside of the academic community, however, his ideas remained influential, as he was a prolific writer featuring often in the mainstream media.

70. For more on Huntington, see G. J. Martin, *Ellsworth Huntington: His Life and Thought* (Hamden, CT: Archon Books, 1973); J. E. Chappell, "Huntington and His Critics: The Influence of Climate on Civilization" (PhD diss., University of Kansas, 1976); J. R. Fleming, "Civilisation, Climate, and Ozone: Ellsworth Huntington's 'Big' Views on Biophysics, Biocosmics, and Biocracy," in *Weather, Climate and the Geographical Imagination*, ed. M. Mahony and S. Randalls (Pittsburgh: University of Pittsburgh Press, 2020), 215–231.

71. E. Huntington, *Civilization and Climate* (New Haven, CT: Yale University Press, 1915).

72. Huntington, *Civilization and Climate*, 150.

73. Huntington, *Civilization and Climate*, 150.

74. Huntington, *Civilization and Climate*, 127; E. Huntington, *World-Power and Evolution* (New Haven, CT: Yale University Press, 1919), 56–57.

75. Huntington, *World-Power and Evolution*, 146.

76. Huntington, *World-Power and Evolution*, 146.

77. Huntington, *World-Power and Evolution*, 146.

78. D. E. Bender, *American Abyss: Savagery and Civilization in the Age of Industry* (Ithaca, NY: Cornell University Press), 60.

79. Bender, *American Abyss*, 40–68.

80. Bender, *American Abyss*, 46.

81. Bender, *American Abyss*, 84–88.

82. S. C. Gilfillan, "The Coldward Course of Progress," *Political Science Quarterly* 35, no. 3 (1920): 393–410.

83. Gilfillan, "Coldward Course of Progress," 410.

84. See M. E. Ackermann, *Cool Comfort: America's Romance with Air-Conditioning* (Washington, DC: Smithsonian Institution Press, 2002), 19.

85. Huntington, *Civilization and Climate*, 285.

86. Huntington, *Civilization and Climate*, 285.

87. Huntington, *Civilization and Climate*, 286.

88. D. Kennedy, "The Perils of the Midday Sun: Climatic Anxieties in the Colonial Tropics," in *Imperialism and the Natural World*, ed. J. M. MacKenzie (Manchester: Manchester University Press, 1990), 118–140.

89. David N. Livingstone has illustrated how this group of thinkers were drawn toward biometeorology that could easily lead to support for eugenic courses. D. N. Livingstone, "Darwinian Hippocratics, Eugenic Enticements, and the Biometeorological Body," 191–214.

90. Martin, *Ellsworth Huntington*, 177.

91. Bender, *American Abyss*, 68.

92. E. Huntington, "Babbitt and the Weather," *Outlook* 153, no. 5 (October 2, 1929): 186.

93. Huntington, "Babbitt and the Weather," 186.

94. Huntington, "Babbitt and the Weather," 186.
95. C. A. Mills, *Living with the Weather* (Cincinnati: Caxton, 1934), 1.
96. Mills, *Living with the Weather*, 2.
97. Mills, *Living with the Weather*, 2.
98. Mills, *Living with the Weather*, 5.
99. Mills, *Living with the Weather*, 28–29.
100. Mills, *Living with the Weather*, 28–29.
101. P. Boyer, *Urban Masses and Moral Order in America* (Cambridge, MA: Harvard University Press, 1978), 220–232.
102. J. Addams, *The Spirit of Youth and the City Streets* (1909; repr., New York: Macmillan, 1912), 51.
103. Addams, *Spirit of Youth*, 4.
104. "Misdirected Energies," *New York Times*, April 2, 1928, 16.
105. "Misdirected Energies."
106. H. H. Heller, "The Playground as a Phase of Social Reform," in *Proceedings of the Second Annual Playground Congress* (New York: Playground Association of America, 1908), 179. Playgrounds were often described this way in the popular press. See, e.g., "The Price of Play," *Chicago Daily Tribune*, March 23, 1913, H4.
107. Heller, "Playground as a Phase," 181.
108. Heller, "Playground as a Phase," 185.
109. "Crime Is Laid to Energies 'Misdirected,'" *Washington Post*, May 14, 1938, x17.
110. "Crime Is Laid," x17.
111. "Crime Is Laid," x17.
112. C. A. Mills, "They Mature *Later* in the Tropics," *Harper's Magazine* 185 (February 1942): 294–298.
113. Mills, "They Mature *Later*," 298.
114. Mills, "They Mature *Later*," 298.
115. Mills, "They Mature *Later*," 297.
116. Mills, "They Mature *Later*," 297.
117. Mills, *Living with the Weather*, 55–59.
118. Mills, *Living with the Weather*, 58.
119. The energetic basis of existence and the dangers of an overstimulating climate discussed by Mills drew on George Miller Beard's diagnosis of neurasthenia. Beard had discussed the peculiarity of the American environment, which drove the American people too fast and to exhaustion. The solutions, Mills proposed, were the same as those adopted for the medical diagnosis of neurasthenia. This included the antidote of outdoor life, hiking, hunting, and bicycling. See D. G. Schuster, *Neurasthenic Nation: America's Search for Health, Happiness, and Comfort, 1869–1920* (New Brunswick, NJ: Rutgers University Press, 2011).
120. G. Gerstle, *American Crucible: Race and Nation in Twentieth Century America* (Princeton, NJ: Princeton University Press, 2001), 95–115.
121. Mills, *Living with the Weather*, 145.
122. Mills, *Living with the Weather*, 149–150.
123. Mills, *Living with the Weather*, 149–150.
124. D. W. Rowland, "Man and Climate," *Ecology* 24, no. 4 (September 1, 1943): 506.

125. See, e.g., "Heat Belt Seen Shifting Northward by Professor," *Evening Independent* (St. Petersburg, Florida), July 28, 1936, 1.

126. See S. P. Weart, *The Discovery of Global Warming* (Cambridge, MA: Harvard University Press, 2003), 3.

127. J. B. Kincer, "Is Our Climate Changing to Milder?," *Scientific Monthly* 39, no. 1 (July 1934): 59–62; "Warmer World," *Time* 33, no. 1 (January 2, 1939): 27.

128. It is important to note that this discussion was not connected to the issue of anthropogenic climate change, caused by rising CO_2 levels, which didn't feature seriously on the public agenda until the late 1940s and early 1950s. See Fleming, *Historical Perspectives on Climate Change*, 115.

129. "True Winters Are Not What They Used to Be," *New York Times*, January 22, 1939, 98; W. Kaempffert, "The Week in Science: A Warmer World?," *New York Times*, December 18, 1938, 63; "Gran'pa's Right! The Record Proves That Winters Are Getting Warmer," *Atlanta Constitution*, April 20, 1939, 1; "World Climate Growing Warmer, Says Russians, Citing Arctic Data," *New York Times*, December 12, 1938, 1.

130. "Workers in the Tropics Are More Efficient," *Daily Boston Globe*, December 27, 1938, 8.

131. E. K. Titus, "Freak Weather Seen Sign of Grave Climatic Changes," *Brooklyn Daily Eagle*, January 8, 1928, B3.

132. Titus, "Freak Weather," B3.

133. Titus, "Freak Weather," B3.

134. T. Hoke, "Utopia by Thermometer," *North American Review* 234, no. 2 (August 1932): 110–116.

135. Hoke, "Utopia by Thermometer," 115.

136. Hoke, "Utopia by Thermometer," 115.

137. Hoke, "Utopia by Thermometer," 110.

138. Hoke, "Utopia by Thermometer," 116.

139. Hoke, "Utopia by Thermometer," 116.

140. C. A. Mills, "Oncoming Reversal of the Human Growth Tide," *Science* 92, no. 2392 (1940): 401–402.

141. "Ebbing Tide," *Time* 36, no. 20 (1940): 43.

142. Rowland, "Man and Climate," 506.

Chapter 3 · *Energizing Culture*

1. L. Mumford, "Personal Note," 1918, box 191, Lewis Mumford Papers, Rare Book and Manuscript Library, University of Pennsylvania (hereafter cited as Mumford Papers).

2. C. Wilson, *The Quest for Wilhelm Reich* (Garden City, NY: Anchor Press / Doubleday, 1981), 212.

3. C. Turner, *Adventures in the Orgasmatron: The Invention of Sex* (London: Fourth Estate, 2012), 5.

4. Reich made the argument that psychoanalysis was conducive to Marxism and its method of dialectical materialism. See W. Reich, "Dialectical Materialism and Psychoanalysis" (1929), in *Sex-Pol*, ed. Lee Baxandall (New York: Random House, 1966), 1–75.

5. W. Reich, *The Function of the Orgasm: Sex-Economic Problems of Biological Energy* (New York: Pocket Books), 17–18.

6. L. Mumford, "Power and Culture," *Third World Power Conference Transactions* (Washington, DC: US Government Printing Office, 1936), 1:167–172.

7. Patrick Geddes coined this term. See P. Geddes, *Cities in Evolution: An Introduction to the Town Planning Movement and to the Study of Civics* (New York: H. Fertig, 1968).

8. Mumford, "Power and Culture," 70.

9. See Turner, *Adventures in the Orgasmatron*, 244–271, 429–446.

10. Mumford, "Power and Culture," 167.

11. Mumford, "Power and Culture," 173.

12. See B. Johnson, *Carbon Nation: Fossil Fuels and the Making of American Culture* (Lawrence: University Press of Kansas, 2014), 105–131; R. C. Tobey, *Technology as Freedom: The New Deal and the Electrical Modernization of the American Home* (Berkeley: University of California Press, 1996); J. Lieberman, *Power Lines: Electricity in American Life and Letters, 1882–1952* (Cambridge, MA: MIT Press, 2017).

13. G. Smith to L. Mumford (n.d.), box 61, Mumford Papers.

14. For a history of the RPAA, see E. K. Spann, *Designing Modern America: The Regional Planning Association of America and Its Members* (Columbus: Ohio State University Press, 1996).

15. Clarence Stein coined this term in a special issue of the *Survey Graphic*. See C. Stein, "Dinosaur Cities," *Survey Graphic* 54, no. 3 (May 1, 1925): 134–138.

16. For more on regionalism, see R. L. Dorman, *Revolt of the Provinces: The Regionalist Movement in America, 1920–1945* (Chapel Hill: University of North Carolina Press, 1993).

17. Smith to Mumford (n.d.).

18. The influence of Freud on Mumford has been mentioned by several commentators. Gregory Morgan Swer has gone furthest in arguing that Mumford's thought was heavily informed by psychoanalysis. See M. Swer, "Technics and (Para)praxis: The Freudian Dimensions of Lewis Mumford's Theories of Technology," *History of Human Sciences* 17, no. 4 (2004): 45–68. See also L. Marx, "Lewis Mumford: Prophet of Organicism," in *Lewis Mumford: Public Intellectual*, ed. T. P. Hughes and A. C. Hughes (Oxford: Oxford University Press, 1990), 164–180; C. Molesworth, "Inner and Outer: The Axiology of Lewis Mumford," in *Lewis Mumford: Public Intellectual*, ed. T. P. Hughes and A. C. Hughes (Oxford: Oxford University Press, 1990), 241–255.

19. In Mumford's working notes there is a record of the psychoanalytic literature he engaged with. This included, among others, E. Jones, "Freud's Psychology," *Bulletin of Psychoanalysis* (April 1910); and P. Janet, "Psychoanalysis," *Journal of Abnormal Psychology* (April–May and June–September 1914). Although undated, these can be found within his notes from 1918–1919; box 191, Mumford Papers.

20. L. Mumford, *The Golden Day: A Study in American Experience and Culture* (New York: Horace Liveright, 1926), 267.

21. For more on the influence of psychoanalysis on American culture, see N. G. Hale, *The Rise and Crisis of Psychoanalysis in the United States: Freud and the Americans, 1917–1985* (Oxford: Oxford University Press, 1995); N. G. Hale, "From Bergstrasse XIX to Central Park West: The Americanization of Psychoanalysis 1919–1940," *Journal of*

the *History of the Behavioral Sciences* 14 (1974): 299–315; L. R. Samuel, *Shrink: A Cultural History of Psychoanalysis in America* (Lincoln: University of Nebraska Press, 2013); J. C. Burnham, "From Avant-Garde to Specialism: Psychoanalysis in America," *Journal of the History of Behavioral Sciences* 15, no. 2 (1979): 128–134.

22. Hale, *Rise and Crisis of Psychoanalysis*, 59.
23. Burnham, "From Avant-Garde to Specialism," 129.
24. See Reich, "Dialectical Materialism and Psychoanalysis," 1–75.
25. Reich, *Function of the Orgasm*, 8.
26. See W. Reich and T. P. Wolfe, *The Mass Psychology of Fascism* (1933; repr., New York: Orgone Institute Press, 1946).
27. S. Freud, "Beyond the Pleasure Principle," in *The Standard Edition of the Complete Psychological Works of Sigmund Freud*, vol. 18, *(1920–1922): Beyond the Pleasure Principle, Group Psychology, and Other Works*, ed. J. Strachey et al. (London: Hogarth, 1955), 7–64. Freud explains how a moral determination is built into the psyche in "The Ego and the Id." He writes, "It may be said of the id that it is totally non-moral, of the ego that it strives to be moral, and of the super-ego that it can be super-moral and then become as cruel as only the id can be." S. Freud, "The Ego and the Id," in *The Standard Edition of the Complete Psychological Works of Sigmund Freud*, vol. 19, *(1923–1925): The Ego and the Id, and Other Works*, ed. J. Strachey et al. (London: Hogarth, 1962), 54.
28. S. Freud, *Civilization and Its Discontents*, in *The Standard Edition of the Complete Psychological Works of Sigmund Freud*, vol. 21, *(1927–1931)*, ed. J. Strachey et al. (London: Hogarth, 1961), 59–145.
29. Turner, *Adventures in the Orgasmatron*, 120.
30. Freud, "Ego and the Id," 12–66.
31. J. L. Thomas, "The Uses of Catastrophism: Lewis Mumford, Vernon L. Parrington, Van Wyck Brooks, and the End of American Regionalism," *American Quarterly* 2, no. 42 (1990): 223–251.
32. Swer, "Technics and (Para)praxis," 53.
33. Swer, "Technics and (Para)praxis," 54.
34. R. Overy, *The Morbid Age: Britain between the Wars* (London: Allen Lane, 2009), 32.
35. O. Spengler, *The Decline of the West*, vol. 2, ed. C. A. Atkinson (New York: A. A. Knopf, 1926).
36. Spengler, *Decline of the West*, 2:392n1.
37. Because of Spengler's association with "Junkerism," the RPAA had a problematic engagement with his thought. In light of this, they reinterpreted Spengler for an American audience through finding moments of reconciliation, purposefully eliminating his Prussian invocations and obfuscation. For an explanation of this, see Thomas, "Uses of Catastrophism." See also D. L. Miller, *Lewis Mumford: A Life* (New York: Weidenfeld & Nicolson, 1989), 300–302.
38. See B. MacKaye, "End or Peak of Civilization?," *Survey Graphic* 68, no. 13 (October 1, 1932): 441.
39. MacKaye, "End or Peak of Civilization?," 444.
40. L. Mumford, "Spengler: Dithyramb to Doom" (1939–1944), in *Interpretations and Forecasts, 1922–1972* (New York: Harcourt Brace Jovanovich, 1973), 221.
41. Mumford, "Spengler," 221.

42. Mumford, "Spengler," 223.

43. See A. Bramwell, *Ecology in the 20th Century: A History* (New Haven, CT: Yale University Press, 1989).

44. See D. Matless, *Landscape and Englishness* (London: Reaktion Books, 1998); D. Matless, "Bodies Made of Grass Made of Earth Made of Bodies: Organicism, Diet and National Health in Mid-twentieth-century England," *Journal of Historical Geography* 27, no. 3 (2001): 355–376.

45. E. Haeckel, quoted in D. Gasman, *The Scientific Origins of National Socialism: Social Darwinism in Ernst Haeckel and the German Monist League* (London: Macdonald, 1971), 69.

46. F. Trentmann, "Civilization and Its Discontents: English Neo-Romanticism and the Transformation of Anti-modernism in Twentieth-Century Western Culture," *Journal of Contemporary History* 29, no. 4 (1994): 602.

47. Trentmann, "Civilization and Its Discontents," 602.

48. One of the most popular books to support this was *Man and Sunlight*, published in England in 1927 and written by Hans Surén, an advocate of the German *Naturkultur* movement. H. Surén, D. A. Jones, and C. W. Saleeby, *Man and Sunlight* (Slough: Sollux, 1927).

49. "The Ceremony of Al-Thing, Bode Bearer 1925, KK VI," box 7, Youth Movement Archive, London School of Economics.

50. D. H. Lawrence, "The Sun" (1925/1928), in *The Cambridge Edition of the Works of D. H. Lawrence: The Woman Who Rode Away and Other Stories*, ed. D. Mehl and C. Jansohn (Cambridge: Cambridge University Press, 1995), 21.

51. D. H. Lawrence, *Studies in Classic American Literature*, ed. E. Greenspan, L. Vasey, and J. Worthen (1923; repr., Cambridge: Cambridge University Press, 2003).

52. Lawrence, *Studies in Classic American Literature*, 102.

53. M. Jefferies and M. Tyldesley, *Rolf Gardiner: Folk, Nature and Culture in Interwar Britain* (Farnham: Ashgate, 2011); R. Gardiner, *World without End: British Politics and the Younger Generation* (London: Cobden-Sanderson, 1932).

54. Gardiner, *World without End*, 15.

55. Gardiner, *World without End*, 35.

56. Gardiner, *World without End*, 38.

57. B. MacKaye to L. Mumford, September 17, 1924, box 37, Mumford Papers.

58. MacKaye to Mumford, September 17, 1924.

59. L. Mumford to B. MacKaye, July 28, 1927, box 79, Mumford Papers.

60. Mumford to MacKaye, July 28, 1927.

61. James E. Strick explains how the energy principle utilized by Freud and Reich drew on hydraulic metaphors. J. E. Strick, *Wilhelm Reich, Biologist* (Cambridge, MA: Harvard University Press, 2015), 13.

62. William James's paradigm of the "stream of consciousness" epitomized this use of topographic metaphor. W. James, *The Principles of Psychology* (1890; repr., Cambridge, MA: Harvard University Press, 1981), 1:246.

63. W. Reich, "The Bioelectrical Function of Sexuality" (unpublished manuscript, May 1938), quoted in Strick, *Wilhelm Reich, Biologist*, 13.

64. For more on the connection between psychology and Tansley's environmental outlook, see P. Anker, *Imperial Ecology: Environmental Order in the British Empire*,

1895–1945 (Cambridge, MA: Harvard University Press, 2001), 23; P. G. Ayres, *Shaping Ecology: The Life of Arthur Tansley* (Chichester: Wiley-Blackwell, 2012), 112; L. Cameron and J. Forrester, "'A Nice Type of English Scientist': Tansley and Freud," *History Workshop Journal* 48 (1999): 65–100.

65. A. Tansley, *The New Psychology and Its Relation to Life* (London: George Allen & Unwin, 1920), 65.

66. Tansley, *New Psychology*, 54.

67. Tansley, *New Psychology*, 64.

68. Tansley, *New Psychology*, 64.

69. Tansley, *New Psychology*, 65. Carl Gustav Jung also utilized these hydraulic metaphors, describing complexes through the valley metaphor. He described how energy flowed down a system in the psyche in the same way as water flowed into a reservoir. C. G. Jung, "On Psychic Energy," in *The Collected Works of C. G. Jung: The Structure and Dynamics of the Psyche*, ed. R. F. C. Hull (London: Routledge & Kegan Paul, 1960), 3–67.

70. Tansley, *New Psychology*, 64–65, 72.

71. Tansley, *New Psychology*, 97.

72. Tansley, *New Psychology*, 97.

73. Mumford quoted an extended section from Tansley's book in a personal note in 1921. "Personal Note" (n.d. [1921]), box 191, Mumford Papers.

74. For more on the relationship between Lewis Mumford and Patrick Geddes, see F. G. Novak Jr., ed., *Lewis Mumford and Patrick Geddes: The Correspondence* (New York: Routledge, 1995).

75. For more on Patrick Geddes, see V. Welter, *Biopolis: Patrick Geddes and the City of Life* (Cambridge, MA: MIT Press, 2002); H. E. Meller, *Patrick Geddes: Social Evolutionist and City Planner* (London: Routledge, 1990).

76. Given that Tansley was associated with the School of Sociology, Mumford is likely to have become aware of Tansley's work while in London during this period. See R. J. Scott and J. Bromley, *Envisioning Sociology: Victor Branford, Patrick Geddes, and the Quest for Social Reconstruction* (Albany: State University of New York Press, 2013).

77. Mumford copied out this passage from p. 42 of *The New Psychology* in his personal notes; box 191, Mumford Papers.

78. See Tansley, *New Psychology*, 86–158.

79. Tansley, *New Psychology*, 148.

80. L. Mumford, "The Fourth Migration," *Survey Graphic* 54, no. 4 (May 1925): 133.

81. Mumford, "Fourth Migration," 133.

82. Mumford, "Fourth Migration," 133.

83. Mumford, "Fourth Migration," 133.

84. Bob Johnson has shown how Giant Power was set within a specific narrative about social rejuvenation and the social benefits of public power. See Johnson, *Carbon Nation*, 114–117.

85. J. Christie, "Giant Power: A Progressive Proposal of the Nineteen-Twenties," *Pennsylvania Magazine of History and Biography* 96, no. 4 (1972): 481.

86. Christie, "Giant Power," 493; Spann, *Designing Modern America*, 70–74.

87. For more on electrification in the Soviet Union, see J. Coopersmith, *The Electrification of Russia, 1880–1926* (Ithaca, NY: Cornell University Press, 1992);

J. Coopersmith, "Technology Transfer in Russian Electrification, 1870–1925," *History of Technology* 13 (1991): 214–234.

88. Lenin had written this in a letter to Gleb Krzhizhanovsky, who was at the forefront of the GOELRO plan. See V. Lenin, *Collected Works*, 35:467–468, quoted in Coopersmith, *Electrification of Russia*, 155.

89. Spann, *Designing Modern America*, 68.

90. S. Chase, *One Billion Wild Horses* (New York: League for Economic Democracy, 1930), 24.

91. Chase, *One Billion Wild Horses*, 24.

92. See S. Chase, "Coals to Newcastle," *Survey Graphic* 24, no. 4 (May 1, 1925): 144, 146. See also D. C. Engerman, *Modernization from the Other Shore: American Intellectuals and the Romance of Russian Development* (Cambridge, MA: Harvard University Press, 2003).

93. R. Bruère, "Pandora's Box," *Survey Graphic* 51, no. 11 (March 1, 1924): 558.

94. R. Bruère, "Giant Power—Region Builder," *Survey Graphic* 54, no. 4 (1925): 162.

95. Bruère, "Giant Power," 162.

96. B. MacKaye, "Regional Planning Studies for the R.P.A. of America" (May 12, 1925), box 38, Mumford Papers.

97. MacKaye, "Regional Planning Studies."

98. MacKaye, "Regional Planning Studies."

99. For more on Benton MacKaye's Appalachian Trail, see P. Sutter, "'A Retreat from Profit': Colonization, the Appalachian Trail, and the Social Roots of Benton MacKaye's Wilderness Advocacy," *Environmental History* 4, no. 4 (1999): 553–577; L. Anderson, *Benton MacKaye, Conservationist, Planner, and Creator of the Appalachian Trail* (Baltimore: Johns Hopkins University Press, 2002).

100. B. MacKaye, "An Appalachian Trail: A Project in Regional Planning," *Journal of the American Institute of Architects* 9 (1921): 325–330.

101. MacKaye, "Appalachian Trail."

102. B. MacKaye, "Appalachian Power: Servant or Master?," *Survey Graphic* 51, no. 11 (March 1, 1924): 619.

103. MacKaye, "Appalachian Power," 618.

104. MacKaye, "Appalachian Power," 619.

105. MacKaye, "Appalachian Power," 619.

106. MacKaye, "End or Peak of Civilization?," 444.

107. MacKaye, "End or Peak of Civilization?," 444.

108. B. MacKaye, *The New Exploration: A Philosophy of Regional Planning* (1928; repr., Champaign: University of Illinois Press, 1962), 33.

109. MacKaye, *New Exploration*, 3–15.

110. MacKaye, *New Exploration*, 5, 7.

111. MacKaye, *New Exploration*, 15.

112. MacKaye, *New Exploration*, 135.

113. MacKaye, *New Exploration*, 135–136.

114. MacKaye, *New Exploration*, 137.

115. MacKaye, *New Exploration*, 137.

116. MacKaye, *New Exploration*, 148.

117. P. Sutter, *Driven Wild: How the Fight against Automobiles Launched the Modern Wilderness Movement* (Seattle: University of Washington Press, 2002), 54–100.

118. Leopold protested not against the recreational use of wilderness per se but against modern recreational practices that involved invasive measures such as road building, hotels, holiday homes, the advertising industry, and light pollution. See Sutter, *Driven Wild*, 70.

119. This essay was part of a series of five articles that explored the notion of wilderness in these terms, each published in different venues, such as *Sunset Magazine*, *USFS Bulletin*, *American Forests and Forest Life*, and *Outdoor Life*. A. Leopold, "Wilderness as a Form of Land Use," *Journal of Land and Public Utility Economics* 1, no. 4 (1925): 398.

120. Leopold, "Wilderness as a Form," 403.

121. There is a large body of scholarship discussing the role of the wilderness in American thought. See R. Nash, *Wilderness and the American Mind* (New Haven, CT: Yale University Press, 1967); M. Lewis, *American Wilderness: A New History* (Oxford: Oxford University Press, 2007).

122. The idea of the wilderness as a scenic resource was not a new idea but extended back to the early conservation movement and preservationists such as John Muir.

123. Leopold, "Wilderness as a Form," 401.

124. For more on the Wilderness Society, see Sutter, *Driven Wild*.

125. Leopold, "Wilderness as a Form," 404.

126. MacKaye, *New Exploration*, 204.

127. MacKaye, *New Exploration*, 204.

128. MacKaye, *New Exploration*, 205.

129. "The Wilderness Society Platform," *Living Wilderness* 1 (September 1935): 2.

130. B. MacKaye, "War and Wilderness," *Living Wilderness* 6 (July 1941): 8.

131. MacKaye, "War and Wilderness," 8.

132. MacKaye, "War and Wilderness," 8.

133. B. MacKaye, "Tennessee—Seed of a National Plan," *Survey Graphic* 22, no. 5 (May 1933): 252. See also B. MacKaye, "The Tennessee River Project: First Step in a National Plan," *New York Times*, April 16, 1933, xx3.

134. MacKaye, "Tennessee," 251.

135. MacKaye, "Tennessee," 254.

136. MacKaye, "Tennessee," 254.

137. Spann, *Designing Modern America*, 157.

138. D. Schaffer, "Ideal and Reality in 1930s Regional Planning: The Case of the Tennessee Valley Authority," *Planning Perspectives* 1, no. 1 (1986): 27–44. For more on Benton MacKaye's influence on the TVA, see D. Schaffer, "Benton MacKaye: The TVA Years," *Planning Perspectives* 5 (1990): 5–21.

139. Tennessee Valley Authority Act, May 18, 1933.

140. T. K. McCraw, *Morgan vs. Lilienthal: The Feud within the TVA* (Chicago: Loyola University Press, 1970); Anderson, *Benton MacKaye*, 248–267.

141. B. MacKaye to S. Chase, July 20, 1935, box 38, Mumford Papers.

142. For more on Roosevelt's engagement with the RPAA, see E. K. Spann, "Franklin Delano Roosevelt and the Regional Planning Association of America, 1931–1936," *New York History* 74, no. 2 (1993): 185–200.

143. Spann, "Franklin Delano Roosevelt," 199–200.

144. B. MacKaye, "An Approach to Planning in the Tennessee Valley" (n.d.), box 180, Mumford Papers.

145. Anderson, *Benton MacKaye*, 261–262.

146. Spann, "Franklin Delano Roosevelt," 199–200.

147. *The Scenic Resources of the Tennessee Valley: A Descriptive and Pictorial Inventory*, prepared by the Department of Regional Planning Studies (Washington, DC: US Government Printing Office, 1938).

148. D. Ekbladh, "Meeting the Challenge from Totalitarianism: The Tennessee Valley Authority as a Global Model for Liberal Development, 1933–1945," *International History Review* 32, no. 1 (2010): 52.

149. O. Keun, *A Foreigner Looks at the TVA* (New York: Longmans, Green, 1937), 4.

150. Keun, *Foreigner Looks*, 3.

151. Keun, *Foreigner Looks*, 3.

152. J. Huxley, *TVA, Adventure in Planning* (Surrey: Architectural Press, 1943), 7.

153. Huxley, *TVA, Adventure in Planning*, 7.

154. Huxley, *TVA, Adventure in Planning*, 135.

155. C. Close, J. S. Huxley, and G. B. Barbour, "The Tennessee Valley Project: Discussion," *Geographical Journal* 89, no. 5 (May 1, 1937): 407.

156. These essays have been collected together in B. MacKaye, *From Geography to Geotechnics* (Champaign: University of Illinois Press, 1968).

157. B. MacKaye, "Folkland as Nation Maker," in *From Geography to Geotechnics* (Champaign: University of Illinois Press, 1968), 60. Reprinted from *Survey* 87, no. 1 (January 1, 1951): 14–16.

158. MacKaye, "Folkland as Nation Maker," 60.

159. Huxley, *TVA, Adventure in Planning*, 136.

160. Huxley, *TVA, Adventure in Planning*, 134.

161. Huxley, *TVA, Adventure in Planning*, 136.

162. Huxley, *TVA, Adventure in Planning*, 136.

163. B. MacKaye, "From Continent to Globe," in *From Geography to Geotechnics* (Champaign: University of Illinois Press, 1968), 97. Reprinted from *Survey* 87, no. 5 (May 1951): 215–218.

Chapter 4 · *Erginette's Beauty*

1. J. Erskine, "Erginette: A Future Helen amid the Topless Towers of Technocracy," *Collier's Weekly*, March 25, 1933, 10.

2. Erskine, "Erginette," 10.

3. Erskine, "Erginette," 28.

4. For the most complete histories of the Technocracy movement, see W. E. Akin, *Technocracy and the American Dream: The Technocrat Movement, 1900–1941* (Berkeley: University of California Press, 1977); H. Elsner, *The Technocrats: Prophets of Automation* (Syracuse: Syracuse University Press, 1967).

5. H. Clurman, *Fervent Years* (New York: Hill & Wang, 1957), 107, quoted in W. E. Leuchtenburg, *Franklin D. Roosevelt and the New Deal: 1932–1940* (New York: Harper & Row, 1963), 18.

6. C. N. Edge, "The Measurements of Technology," *Living Age*, January 1, 1933, 400.

7. "Industrial Growth of Nation Is Traced: 'Energy Survey' Undertaken by Columbia University to Cover the Last 100 Years," *New York Times*, August 6, 1932, 13.

8. Leuchtenburg, *Franklin D. Roosevelt*, 23.

9. S. Chase, *The Economy of Abundance* (New York: Macmillan, 1934), 5.

10. Chase, *Economy of Abundance*, 3.

11. G. W. Johnson, "Bryan, Thou Shouldst Be Living," *Harper's Monthly Magazine* 163, no. 976 (September 1931): 388.

12. V. Jordan, "Technocracy—Tempest on a Slide Rule," *Scribner's Magazine* 93, no. 2 (February 1933): 68, 66.

13. Jordan, "Technocracy," 68.

14. J. Martinez-Alier, *Ecological Economics: Energy, Environment, and Society* (Oxford: Basil Blackwell, 1987); P. Mirowski, *More Heat Than Light: Economics as Social Physics and Physics as Nature's Economics* (Cambridge: Cambridge University Press, 2000). See also P. Mirowski, "Energy and Energetics in Economic Theory: A Review Essay," *Journal of Economic Issues* 22, no. 3 (September 1988): 811–830.

15. Histories of Technocracy have tended to represent it as an eccentric movement that only had a brief spell of popularity before it fell out of favor. Akin, *Technocracy and the American Dream*; Elsner, *Technocrats*. In this chapter, however, I will demonstrate that Technocracy had a far wider influence beyond the movement, entering into the cultural imaginary.

16. "Technocracy Idea Borrowed, He Says," *New York Times*, December 30, 1932, 13.

17. H. Scott, "Technology Smashes the Price System," *Harper's Monthly Magazine* 166 (January 1933): 130.

18. Scott, "Technology Smashes the Price System," 131.

19. Scott, "Technology Smashes the Price System," 131.

20. Leuchtenburg, *Franklin D. Roosevelt*, 19.

21. Scott, "Technology Smashes the Price System," 141–142.

22. "Technocracy Idea Borrowed, He Says," 13.

23. F. Soddy, *Cartesian Economics: The Bearing of Physical Science upon State Stewardship* (London: Hendersons, 1922), 29, 28.

24. Soddy, *Cartesian Economics*, 30.

25. I am quoting here from the second edition published in America in 1933, which included a short note on the rise of the Technocracy movement. F. Soddy, *Wealth, Virtual Wealth and Debt: The Solution of the Economic Paradox* (New York: E. P. Dutton, 1933), 10.

26. Soddy, *Wealth, Virtual Wealth and Debt*, 76–77.

27. Soddy, *Wealth, Virtual Wealth and Debt*, 51.

28. The suspicion over money's shibboleth character has a long history. See M. Shell, *Money, Language, and Thought: Literary and Philosophical Economies from the Medieval to the Modern Era* (Berkeley: University of California Press, 1982).

29. A. Director, *The Economics of Technocracy*, ed. Harry D. Gideonse (Chicago: University of Chicago Press, 1933), 5.

30. L. Merricks, *The World Made New: Frederick Soddy, Science, Politics, and Environment* (Oxford: Oxford University Press, 1996), 108. See also J. L. Finlay, *Social Credit: The English Origins* (Montreal: McGill–Queens University Press, 1972).

31. Akin, *Technocracy and the American Dream*, x–xi.

32. T. Veblen and D. Bell, *The Engineers and the Price System* (1921; repr., New York: Harcourt, Brace & World, 1963).

33. For more on the Technical Alliance, see L. Ardzrooni, "Veblen and Technocracy," *Living Age* 344 (March 1933): 39–40; M. Spindler, *Veblen and Modern America: Revolutionary Iconoclast* (London: Pluto, 2002); J. Dorfman, *Thorstein Veblen and His America* (New York: Viking, 1934).

34. "Industrial Growth of Nation," 13.

35. B. J. Eichengreen, *Golden Fetters: The Gold Standard and the Great Depression, 1919–1939* (Oxford: Oxford University Press, 1992).

36. D. M. Kennedy, *Freedom from Fear: The American People in Depression and War, 1929–1945* (Oxford: Oxford University Press, 1999), 35.

37. Kennedy, *Freedom from Fear*, 35.

38. Kennedy, *Freedom from Fear*, 37.

39. I. Fisher, *Stabilised Money: A History of the Movement* (London: George Allen & Unwin, 1935), xv–xvi.

40. Martinez-Alier, *Ecological Economics*.

41. F. Soddy, "Men, Machines, Money—and the Technocrats," *Daily Mail*, January 7, 1933.

42. P. Geddes, *John Ruskin, Economist* (Edinburgh: William Brown, 1884), 26.

43. Geddes, *John Ruskin, Economist*, 26.

44. Martinez-Alier, *Ecological Economics*, 128.

45. Scott, "Technology Smashes the Price System," 132.

46. Geddes, *John Ruskin, Economist*, 26.

47. Scott, "Technology Smashes the Price System," 131–132.

48. Scott, "Technology Smashes the Price System," 132.

49. W. W. Parrish, "Technocracy's Challenge," *New Outlook* 161, no. 4 (January 1933): 14.

50. Director, *Economics of Technocracy*, 7.

51. Director, *Economics of Technocracy*, 7.

52. H. Loeb, *Life in a Technocracy, What It Might Be Like*, ed. H. P. Segal (1933; repr., Syracuse: Syracuse University Press, 1996), 75.

53. I. Katznelson, *Fear Itself* (New York: Liveright, 2013), 105.

54. Katznelson, *Fear Itself*, 107.

55. Kennedy, *Freedom from Fear*, 111.

56. B. Alpers, *Dictators, Democracy, and American Public Culture: Envisioning the Totalitarian Enemy, 1920s–1950s* (Chapel Hill: University of North Carolina Press, 2003), 15–59.

57. R. C. Tugwell, *The Brains Trust* (New York: Viking, 1968), 62, quoted in Kennedy, *Freedom from Fear*, 111.

58. Jordan, "Technocracy," 66.

59. "Vox Populi Hails Technocracy—or Anything Else That'll Help," *Brooklyn Daily Eagle*, January 17, 1933, 3.

60. For more on *Broom*, see P. Nicholls, "Destinations, *Broom* (1921–4) and *Secession* (1922–4)," in *The Oxford Critical and Cultural History of Modernist Magazines*, vol. 2, *North America 1894–1960*, ed. P. Brooker and A. Thacker (Oxford: Oxford University

Press, 2009), 636–654; H. P. Segal, introduction to *Life in a Technocracy*, by H. Loeb (Syracuse: Syracuse University Press, 1996), xv.

61. Loeb, *Life in a Technocracy*, 104.
62. Loeb, *Life in a Technocracy*, 104–109.
63. Loeb, *Life in a Technocracy*, 141–142.
64. Loeb, *Life in a Technocracy*, 140–141.
65. A. MacLeish, "Machines and the Future," *Nation* 136, no. 3527 (February 8, 1933): 142.
66. Jordan, "Technocracy," 66.
67. For more on Hazlitt's debunking of absolute value, see his altercation with the writer James Truslow Adams in H. Hazlitt, "How High Is Up? The Fallacy of Absolute Standards," *Forum* 87, no. 2 (February 1932): 115–119.
68. H. Hazlitt, "Scrambled Ergs: Examination of Technocracy," *Nation* 136, no. 3526 (February 1, 1933): 115.
69. Hazlitt, "Scrambled Eggs," 115.
70. C. Frederick, "The Psychology of Women under Technocracy," in *For and Against, Technocracy, a Symposium*, ed. J. G. Frederick (New York: Business Bourse, 1933), 111.
71. Frederick, "Psychology of Women under Technocracy," 113.
72. Frederick, "Psychology of Women under Technocracy," 121.
73. Frederick, "Psychology of Women under Technocracy," 129.
74. Frederick, "Psychology of Women under Technocracy," 129.
75. The correlation of women to waste fed through narratives of Taylorism and industrial production. Women often symbolized excess within the industrial system. Martha Banta has demonstrated that whereas men came to denote fact, women were often equated with subjective value and desire. See M. Banta, *Taylored Lives: Narrative Productions in the Age of Taylor, Veblen, and Ford* (Chicago: University of Chicago Press, 1993).
76. Frederick, "Psychology of Women under Technocracy," 129.
77. T. Armstrong, "Social Credit Modernism," *Critical Quarterly* 55, no. 2 (2013): 50–65. For outlines of the Social Credit movement, see Finlay, *Social Credit*; F. Hutchinson and B. Burkitt, *The Political Economy of Social Credit and Guild Socialism* (London: Routledge, 1997).
78. For an assessment of Social Credit views in America, see R. Preda, "Social Credit in America: A View from Pound's Economic Correspondence, 1933–1940," *Paideuma* 34, nos. 2 and 3 (2005): 201–231.
79. J. Hargrave, "They Can't Kill the Sun" (n.d.), box 62, Youth Movement Archive, London School of Economics.
80. For more on Hargrave and the Green Shirts, see M. Drakeford and J. Campling, *Social Movements and Their Supporters: The Green Shirts in England* (New York: St. Martin's, 1997).
81. Hargrave, "They Can't Kill the Sun."
82. Hargrave, "They Can't Kill the Sun."
83. Hargrave, "They Can't Kill the Sun."
84. Hargrave, "They Can't Kill the Sun."

85. Hargrave, "They Can't Kill the Sun."
86. Hargrave, "They Can't Kill the Sun."
87. H. Scott to T. Dreiser, January 7, 1936, box 105, folder 6080, Theodore Dreiser Papers, University of Philadelphia, Ms. Coll. 30.
88. Scott to Dreiser, January 7, 1936.
89. Scott to Dreiser, January 7, 1936.
90. Hargrave, "They Can't Kill the Sun."
91. For more on anti-Semitism in the Great Depression, see B. S. Wenger, *New York Jews and the Great Depression: Uncertain Promise* (Syracuse: Syracuse University Press, 1999).
92. This association ran through Brooks Adams's "Law," with money power, Wall Street, and Jewry signaling the last stage before civilization's collapse. See, e.g., B. Adams, *The Law of Civilization and Decay: An Essay on History* (1895; repr., New York: Alfred A. Knopf, 1943), 362.
93. P. Morrison, *The Poetics of Fascism: Ezra Pound, T. S. Eliot, Paul de Man* (New York: Oxford University Press, 1996), 58.
94. Soddy, *Cartesian Economics*, 32.
95. W. Crick and F. Soddy, *Abolish Private Money, or Drown in Debt* (London: Sault, 1939), 5.
96. A. Brinkley, *Voices of Protest: Huey Long, Father Coughlin, and the Great Depression* (New York: Knopf, 1982), 119.
97. For more on Coughlin, see D. I. Warren, *Radio Priest: Charles Coughlin, the Father of Hate Radio* (New York: Free Press, 1996).
98. Brinkley, *Voices of Protest*, 94–94.
99. C. E. Coughlin, "The God of Gold," in *Father Coughlin's Radio Discourses, 1931–32* (Royal Oak, MI: Radio League of the Little Flower, 1932), 161.
100. Coughlin, "God of Gold," 166.
101. C. E. Coughlin, *Money! Questions and Answers* (Royal Oak, MI: National Union for Social Justice, 1936), 11.
102. E. Pound, *"Ezra Pound Speaking": Radio Speeches of WWII*, ed. L. W. Doob (Westport, CT: Greenwood, 1978), 256.
103. See L. Surette, *Pound in Purgatory: From Economic Radicalism to Anti-Semitism* (Champaign: University of Illinois Press, 1999).
104. Morrison, *Poetics of Fascism*, 19.
105. Pound's attraction to fascism and anti-Semitism, Leon Surette has argued, occurred, as it did for many of his generation, through a direct engagement with the economic theories common at the time. See Surette, *Pound in Purgatory*.
106. For Gesell's influence on Pound, see T. Redman, "The Discovery of Gesell," in *Ezra Pound and Italian Fascism* (Cambridge: Cambridge University Press, 1991), 122–155.
107. E. Fenollosa and E. Pound, "The Chinese Written Character as a Medium for Poetry: An Ars Poetica (1918–1936)," in *The Chinese Written Character as a Medium for Poetry: A Critical Edition, Ernest Fenollosa and Ezra Pound*, ed. H. Saussy, J. Stalling, and L. Klein (New York: Fordham University Press, 2008), 41–60.
108. Fenollosa and Pound, "The Chinese Written Character," 58.
109. Fenollosa and Pound, "The Chinese Written Character," 50.

110. F. D. Roosevelt, "Campaign Address on Public Utilities and Development of Hydro-electric Power" (Portland Oregon, September 21, 1932), in *The Public Papers and Addresses of Franklin D. Roosevelt*, vol. 1, *The Genesis of the New Deal* (New York: Random House, 1938), 727–742; J. A. Hagerty, "Portland Cheers Speech," *New York Times*, September 22, 1932, 1.

111. Roosevelt, "Campaign Address," 739.

112. For a discussion of the "yardstick" slogan, see S. M. Neuse, *David E. Lilienthal: The Journey of an American Liberal* (Knoxville: University of Tennessee Press, 1996), 82–86.

113. This language was common. See, e.g., Loeb, *Life in a Technocracy*, 75.

114. W. Parrish, "Technocracy's Question," *New Outlook* 161, no. 3 (December 1932): 14.

115. See R. C. Tobey, *Technology as Freedom: The New Deal and the Electrical Modernization of the American Home* (Berkeley: University of California Press, 1996).

116. Roosevelt, "Campaign Address," 740.

117. This was how it was described by David Lilienthal in his well-known book *TVA—Democracy on the March* (New York: Harper, 1944).

118. "Muscle Shoals as a Yardstick for Electric Rates," *Literary Digest* 116, no. 14 (September 30, 1933): 9.

119. "Muscle Shoals as a Yardstick," 9.

120. "That Rubber Yard," *New York Times*, November 4, 1934, E4; "Rubber on the Tennessee," *New York Times*, May 24, 1935, 20.

121. "Yardsticks," *New York Times*, August 27, 1933, 4.

122. "Yardsticks," 4.

123. "A Yardstick No More," *New York Times*, July 14, 1935, E8; "Head of TVA Denies Utilities' Charges: Lilienthal Insists before House Group That Project Is Not a 'Rubber Yardstick,'" *New York Times*, April 5, 1935, 39.

124. S. Chase, *A New Deal* (New York: Macmillan, 1932), 177.

125. R. B. Westbrook, "Tribune of the Technostructure: The Popular Economics of Stuart Chase," *American Quarterly* 32, no. 4 (1980): 387–408.

126. Chase, *Economy of Abundance*, 5; Westbrook, "Tribune of the Technostructure," 389. Chase did not wholly accept all of Technocracy's proposals. He asked for a "chaser of psychology and anthropology" to fill the gaps in Technocracy's theory of energy determinants. However, although Chase admitted that there were limitations to its theory, he added that this did not mean that their "conclusions are shaky, only that they do not go far enough." S. Chase, *Technocracy: An Interpretation* (New York: John Day Pamphlets, 1933), 32.

127. "Analysis of Senator Watson's Charge against the Federal Trade Commission and Members of the Commission's Staff," box 2, Stuart Chase Papers, Manuscript Division, Library of Congress, Washington, DC (hereafter cited as Chase Papers).

128. S. Chase, "Portrait of a Radical," *Century Magazine* 108, no. 3 (July 1924): 301.

129. Chase, *New Deal*, 177.

130. In the research notes to *The Economy of Abundance* Chase refers to Soddy's book *Wealth, Virtual Wealth and Debt*, pointing to the influence Soddy had on his work. "Research Notes," for *The Economy of Abundance*, circa 1931–1938, box 15, Chase Papers.

131. "(Undated note)," box 15, Chase Papers.

132. Chase, "Portrait of a Radical," 301.
133. Chase, "Portrait of a Radical," 302.
134. Chase, "Portrait of a Radical," 301.
135. Chase, "Portrait of a Radical," 301.
136. Chase, *Economy of Abundance*, 5.
137. Chase, *Economy of Abundance*, 5–6.
138. Chase, *Economy of Abundance*, 6.
139. Chase, *Economy of Abundance*, 6–8.
140. Chase, *Economy of Abundance*, 10.
141. Chase, *Economy of Abundance*, 10.
142. S. Chase, *Mexico: A Study of Two Americas* (London: John Lane the Bodley Head, 1932), 15–18.
143. Chase, *Mexico*, 223.
144. Chase, *Mexico*, 223.
145. Chase, *Mexico*, 17, 223–224.
146. Chase, *Mexico*, 223.
147. Chase, *Mexico*, 223.
148. For more on Norris Village, see W. L. Creese, *TVA's Public Planning: The Vision, the Reality* (Knoxville: University of Tennessee Press, 1990), 239–324.
149. For more on the influence of the RPAA on the TVA plan, see Creese, *TVA's Public Planning*, 42–52, 57, 248; E. K. Spann, "Franklin Delano Roosevelt and the Regional Planning Association of America, 1931–1936," *New York History* 74, no. 2 (1993): 185–200.
150. Creese, *TVA's Public Planning*, 243.
151. "Technocrat Gives 'Energy' as a Fee," *New York Times*, January 9, 1933, 28; "One of Technocracy's Electricity Dollars Is Taken in Collection," *Ottawa Citizen*, January 9, 1933, 1.
152. The relationship between Technocracy and Christianity was debated in church halls, in public meetings, and at different society groups across the country. See, e.g., "Sees Christianity, Not Technocracy, as Cure," *Dayton Daily News*, January 17, 1933, 9; E. H. J. Hoh, "Christianity and Technocracy," *Mercury*, January 31, 1933, 2.
153. P. W. Ramsey, "Technocracy, 'Energy Units' as a Basis for Exchange Fail to Give Proper Consideration to 'Human Element' Says Tech Economics Head," *Pittsburgh Post-Gazette*, December 21, 1932, 4.
154. Ramsey, "Technocracy, 'Energy Units,'" 4.
155. "Technocracy Like Dream of Ostrich, House Told," *Los Angeles Times*, January 15, 1933, 10.
156. W. Lippmann, "Technocracy: The Prophecy of Doom," *Boston Globe*, February 2, 1933, 14.
157. Lippmann, "Technocracy," 6.
158. Hazlitt, "Scrambled Ergs," 115.
159. "Money Systems on Technocracy Weak Aver Economists," *Ellensburg Daily Record*, December 31, 1932, 1–2.
160. "Money Systems on Technocracy Weak," 1–2.
161. G. Soule, "A Critique of Technocracy's Five Main Points," in *For and Against Technocracy: A Symposium*, ed. J. G. Frederick (New York: Business Bourse, 1933), 104.

162. Soule, "Critique," 104.

163. F. A. Hayek, *The Counter-revolution of Science: Studies on the Abuse of Reason* (Glencoe, IL: Free Press, 1952), 51. For more on Hayek's critique of the energy theory of value, see J. O'Neill, "Ecological Economics and the Politics of Knowledge: The Debate between Hayek and Neurath," *Cambridge Journal of Economics* 28, no. 3 (May 1, 2004): 431–447.

164. J. Massey, "Necessary Beauty: Fuller's Sumptuary Aesthetics," in *New Views on R. Buckminster Fuller*, ed. H. Chu and R. G. Trujillo (Stanford, CA: Stanford University Press, 2009), 109.

165. R. B. Fuller, *Nine Chains to the Moon* (New York: Lippincott, 1938), 67.

166. Fuller, *Nine Chains to the Moon*, 18.

167. R. B. Fuller, *Ideas and Integrities: A Spontaneous Autobiographical Disclosure* (Englewood Cliffs, NJ: Prentice Hall, 1963), 22, 23.

168. For more on how industrial and military vocabulary influenced Fuller, see P. Anker, "Buckminster Fuller as Captain of Spaceship Earth," *Minerva* 45, no. 4 (2007): 417–434.

169. Fuller, *Ideas and Integrities*, 25.

170. Fuller, *Ideas and Integrities*, 25.

171. Fuller, *Nine Chains to the Moon*, 70.

172. Fuller, *Nine Chains to the Moon*, 70.

173. Fuller, *Nine Chains to the Moon*, 73.

174. Fuller, *Nine Chains to the Moon*, 74.

175. Fuller, *Nine Chains to the Moon*, 82.

176. Fuller, *Nine Chains to the Moon*, 91.

177. Fuller, *Nine Chains to the Moon*, 93.

178. Fuller, *Nine Chains to the Moon*, 93.

179. Fuller, *Nine Chains to the Moon*, 102.

180. Fuller, *Nine Chains to the Moon*, 102.

181. Fuller, *Nine Chains to the Moon*, 104.

182. Fuller, *Nine Chains to the Moon*, 104.

183. Fuller, *Nine Chains to the Moon*, 82.

184. S. D. Cox, *The Woman and the Dynamo: Isabel Paterson and the Idea of America* (New Brunswick, NJ: Transaction, 2004), 178.

185. I. Paterson, "Turns with a Bookworm," *New York Herald Tribune "Books,"* September 1, 1933, 11.

186. For more on their relationship, see J. Burns, *Goddess of the Market: Ayn Rand and the American Right* (Oxford: Oxford University Press, 2009), 74–78.

187. Cox, *Woman and the Dynamo*, 142.

188. Cox, *Woman and the Dynamo*, 140–141.

189. I. Paterson, *The God of the Machine* (New York: G. P. Putnam's Sons, 1943), 32.

190. I. Paterson, "Books," quoted in *For and Against Technocracy, A Symposium*, ed. J. G. Frederick (New York: Business Bourse, 1933), 275.

191. Paterson, "Books," 275.

192. Paterson, "Books," 275.

193. Paterson, "Books," 276.

194. Paterson, "Books," 276.

195. Cox, *Woman and the Dynamo*, 256–257.
196. Paterson, *God of the Machine*, 13.
197. Paterson, *God of the Machine*, 31–32.
198. I. Paterson, "Turns with a Bookworm," *Herald Tribune Book Review*, July 16, 1939, quoted in Cox, *Woman and the Dynamo*, 257.
199. Paterson, *God of the Machine*, 151–152.
200. Paterson, *God of the Machine*, 152.
201. Paterson, *God of the Machine*, 152.
202. Paterson, *God of the Machine*, 280.
203. Paterson, *God of the Machine*, 280–281.
204. Paterson, *God of the Machine*, 281.
205. J. Storck, "The Dynamics of Our Society: *The God of the Machine*, by Isabel Paterson," *New York Times*, May 23, 1943, 29.
206. Storck, "Dynamics of Our Society," 29.
207. Storck, "Dynamics of Our Society," 29.
208. Storck, "Dynamics of Our Society," 29.
209. H. Frazer, "The Energy Certificate," *Technocracy Inc.* A, no. 10 (July 1937): 11–13.

Chapter 5 · Energizing the World

1. A. Huxley, *After Many a Summer Dies the Swan* (1939; repr., Chicago: Ivan R. Dee, 1993), 149.
2. N. Cousins, "Modern Man Is Obsolete," *Saturday Review* 28, no. 32 (August 18, 1945): 5.
3. Cousins, "Modern Man Is Obsolete," 5.
4. Cousins, "Modern Man Is Obsolete," 4.
5. Cousins, "Modern Man Is Obsolete," 8.
6. P. S. Boyer, *By the Bomb's Early Light: American Thought and Culture at the Dawn of the Atomic Age* (New York: Pantheon, 1985), 40.
7. See, e.g., W. E. Hocking, "The Atom as Moral Dictator," *Saturday Review of Literature* 29, no. 5 (February 2, 1946): 7–8; E. Lindeman, "Morality for the Atomic Age," *Forum* 106, no. 3 (November 1, 1945): 231–233.
8. "Ethics for an Atomic Age," *Saturday Review* 28, no. 35 (September 1, 1945): 18.
9. See T. H. Huxley, "Evolution and Ethics" (1893), in *Evolution and Ethics and Other Essays* (London: Macmillan, 1895).
10. "Ethics for an Atomic Age," 18.
11. T. H. Huxley, "Evolution and Ethics: Prolegomena" (1894), in *Evolution and Ethics and Other Essays* (London: Macmillan, 1895), 8.
12. Although Cousins's article cited Julian Huxley as a point of reference, it is important to note how heavily indebted Cousins's argument was to Julian Huxley's "scientific humanism." In fact, "Modern Man Is Obsolete" rehearsed the central argument of an article published by Julian Huxley in *Science*, titled "Science, War and Reconstruction." See J. Huxley, "Science, War and Reconstruction," *Science* 91, no. 2355 (1940): 151–158.
13. J. Huxley, *Evolutionary Ethics* (Oxford: Oxford University Press, 1943), 9.
14. Huxley, *Evolutionary Ethics*, 9.
15. Huxley, *Evolutionary Ethics*, 67.

16. See Huxley, *Evolutionary Ethics*, 63–64.

17. There is extensive scholarship on the atomic consciousness that emerged after the Second World War. See Boyer, *By the Bomb's Early Light*; A. Nadel, *Containment Culture: American Narrative, Postmodernism, and the Atomic Age* (Durham, NC: Duke University Press, 1995); M. A. Henriksen, *Dr. Strangelove's America: Society and Culture in the Atomic Age* (Berkeley: University of California Press, 1997); M. Scheibach, *Atomic Narratives and American Youth: Coming of Age with the Atom, 1945–1955* (Jefferson, NC: McFarland, 2003).

18. D. Ciepley, *Liberalism in the Shadow of Totalitarianism* (Cambridge, MA: Harvard University Press, 2006), 1. For more on the perception of dictatorship in American popular culture, see B. L. Alpers, *Dictators, Democracy, and American Public Culture: Envisioning the Totalitarian Enemy, 1920–1950s* (Chapel Hill: University of North Carolina Press, 2003).

19. Ciepley, *Liberalism in the Shadow*, 21–22.

20. K. Mattson, *Intellectuals in Action: The Origins of the New Left and Radical Liberalism, 1945–1970* (University Park: Pennsylvania State University Press, 2002), 30–36.

21. For more on this shift, see Mattson, *Intellectuals in Action*; R. Genter, *Late Modernism: Art, Culture, and Politics in Cold War America* (Philadelphia: University of Pennsylvania Press, 2010).

22. For more on the development of Mumford's thought during this period, see R. W. Fox, "Tragedy, Responsibility, and the American Intellectual, 1925–1950," in *Lewis Mumford: Public Intellectual*, ed. T. P. Hughes and A. C. Hughes (Oxford: Oxford University Press, 1990), 323–337; E. Mendelsohn, "Prophet of Our Discontent: Lewis Mumford Confronts the Bomb," in *Lewis Mumford: Public Intellectual*, ed. T. P. Hughes and A. C. Hughes (Oxford: Oxford University Press, 1990), 343–360. Mark Grief has pointed out how planning rhetoric was often translated into discourse on the crisis of man. See M. Greif, *The Age of the Crisis of Man: Thought and Fiction in America, 1933–1973* (Princeton, NJ: Princeton University Press, 2015), 63.

23. For more on deadly orgone, see C. Turner, *Adventures in the Orgasmatron* (London: Harper, 2008), 331–336.

24. D. L. Miller, *Lewis Mumford: A Life* (New York: Weidenfeld & Nicolson, 1989), 424–425.

25. L. Mumford, *The Conduct of Life* (New York: Harcourt, Brace, 1951), 9.

26. Mumford, *Conduct of Life*, 3.

27. Mumford, *Conduct of Life*, 13.

28. Mumford, *Conduct of Life*, 13.

29. Mumford, *Conduct of Life*, 16.

30. Mumford, *Conduct of Life*, 17.

31. Miller, *Lewis Mumford*, 394.

32. L. Mumford, "The Corruption of Liberalism," *New Republic*, April 29, 1940, 569, 572. Mumford's article was one of many to attack pragmatism for undermining America's resistance to fascism. See W. Frank, "Our Guilt in Fascism," *New Republic* 102 (May 6, 1940): 603–608; A. MacLeish, "The Irresponsibles," *Nation* 150 (May 18, 1940): 618–623.

33. Mumford, "Corruption of Liberalism," 570.

34. Mumford, "Corruption of Liberalism," 573.
35. Miller, *Lewis Mumford*, 415.
36. See the chapter "Cosmos and Person" in Mumford, *Conduct of Life*, 58–91.
37. L. Mumford, *The Condition of Man* (New York: Harcourt, Brace, 1944), 8.
38. Mumford, *Condition of Man*, 202.
39. Mumford, *Condition of Man*, 202.
40. Mumford, *Condition of Man*, 201.
41. Mumford, *Conduct of Life*, 10.
42. Miller, *Lewis Mumford*, 416.
43. Toynbee appeared on the cover of *Time* magazine on March 17, 1947. [Cover], *Time* 49, no. 11 (March 17, 1947).
44. A. Toynbee, "Does History Always Repeat Itself? Professor Toynbee, Examining the Rise and Fall of Civilizations, Finds Hope for Ours," *New York Times*, September 21, 1947, SM15.
45. Toynbee, "Does History Always Repeat Itself?," SM15. It is important to note that Toynbee was influenced by Ellsworth Huntington, especially the connection he made between human beings and their environment. See G. J. Martin, *American Geography and Geographers: Toward Geographical Science* (Oxford: Oxford University Press, 2015), 404–405.
46. Toynbee, "Does History Always Repeat Itself?," SM15.
47. A. Toynbee, *A Study of History*, ed. D. C. Somervell (Oxford: Oxford University Press, 1947), 39.
48. Toynbee, *Study of History*, 189.
49. Toynbee, *Study of History*, 198.
50. Toynbee, *Study of History*, 198.
51. Toynbee, *Study of History*, 364–365.
52. J. Campbell, *The Hero with a Thousand Faces* (1949; repr., Princeton, NJ: Princeton University Press, 1972), 30.
53. Campbell, *Hero with a Thousand Faces*, 36.
54. Campbell, *Hero with a Thousand Faces*, 316, 181.
55. Campbell, *Hero with a Thousand Faces*, 189, 342.
56. Campbell, *Hero with a Thousand Faces*, 40.
57. Campbell, *Hero with a Thousand Faces*, 391.
58. T. Clark, *Charles Olson: The Allegory of a Poet's Life* (New York: Norton, 1991), 76–89.
59. C. Olson, "Projective Verse" (1950), in *Collected Prose: Charles Olson*, ed. D. Allen and B. Friedlander (Berkeley: University of California Press, 1997), 40.
60. C. Olson, "The Gate and the Center" (1951), in *Collected Prose: Charles Olson*, ed. D. Allen and B. Friedlander (Berkeley: University of California Press, 1997), 170.
61. Olson, "Gate and the Center," 172.
62. Olson, "Gate and the Center," 172.
63. Olson, "Gate and the Center," 172–173.
64. H. Rosenberg, "The American Action Painters," *Art News* 51, no. 8 (December 1952). Reprinted in H. Rosenberg, *The Tradition of the New* (New York: Horizon, 1959), 23–39. Byers discusses the shared practices between Olson and the abstract

expressionists. See M. Byers, *Charles Olson and American Modernism: The Practice of the Self* (Oxford: University of Oxford Press, 2018).

65. Rosenberg, "American Action Painters," 29.

66. Rosenberg, "American Action Painters," 25.

67. Rosenberg, "American Action Painters," 29.

68. See D. Belgrad, *The Culture of Spontaneity: Improvisation and the Arts in Postwar America* (Chicago: University of Chicago Press, 1999), 120–140; R. Motherwell, April 1979, quoted in Belgrad, *Culture of Spontaneity*, 122; J. Pollock, "Handwritten Statement" (n.d.), quoted in P. Karmel, *Jackson Pollock: Interviews, Articles, and Reviews* (New York: Museum of Modern Art, 1999), 24.

69. "Jackson Pollock, Is He the Greatest Living Painter in the United States?," *Life* 27, no. 6 (August 8, 1949): 42–45; "The Champ," *Time* 66, no. 25 (December 19, 1955), 66, quoted in Genter, *Late Modernism*, 208.

70. S. Sillen, "Lewis Mumford's 'Mein Kampf,'" *New Masses*, October 15, 1940, 8.

71. Sillen, "Lewis Mumford's 'Mein Kampf,'" 8.

72. Sillen, "Lewis Mumford's 'Mein Kampf,'" 8.

73. Sillen, "Lewis Mumford's 'Mein Kampf,'" 8.

74. See S. Hook, "The New Failure of Nerve," *Partisan Review* 10, no. 1 (January–February 1943): 2–23. For a full account of the debate around the "New Failure of Nerve," see Byers, *Charles Olson and American Modernism*, 20–44.

75. Hook, "New Failure of Nerve," 9.

76. Hook, "New Failure of Nerve," 9.

77. Hook, "New Failure of Nerve," 2.

78. Stuart Chase often described the Soviet Union in these terms. See S. Chase, *One Billion Wild Horses* (New York: League for Economic Democracy, 1930).

79. This language was also fostered within the Soviet Union. See H. Proctor, "Reason Displaces All Love: Libidinal Economizing in the NEP-Era Soviet Union," *New Inquiry*, February 2014.

80. Dewey wrote about his experiences in a series of articles published in the *New Republic* in November and December 1928. This was later published as J. Dewey, *Impressions of Soviet Russia and the Revolutionary World: Mexico—China—Turkey* (1929; repr., New York: Teachers College, Columbia University, 1956), 62.

81. Dewey, *Impressions of Soviet Russia*, 62–63.

82. Dewey, *Impressions of Soviet Russia*, 63–64.

83. For more on the popular representation of Russia during this period, see P. G. Filene, *Americans and the Soviet Experiment, 1917–1933* (Cambridge, MA: Harvard University Press, 1967), 187–241.

84. L. Fischer, "Russia's New Revolution," *Nation* 130, no. 3376 (March 19, 1930): 322.

85. Fischer, "Russia's New Revolution," 322.

86. Fischer, "Russia's New Revolution," 323.

87. For more on Dewey's relationship to the Soviet Union, see R. B. Westbrook, *John Dewey and American Democracy* (Ithaca, NY: Cornell University Press, 1991), 463–495. See also D. C. Engerman, "John Dewey and the Soviet Union: Pragmatism Meets Revolution," *Modern Intellectual History* 3, no. 1 (2006): 33–63.

88. J. Burns, *Goddess of the Market: Ayn Rand and the American Right* (Oxford: Oxford University Press, 2009), 45.

89. A. Rand, *The Fountainhead* (London: Penguin Books, 1994), 343.

90. Rand, *Fountainhead*, 711.

91. Rand, *Fountainhead*, 370.

92. Rand, *Fountainhead*, 668.

93. N. G. Hale, *The Rise and Crisis of Psychoanalysis in the United States: Freud and the Americans, 1917–1985* (New York: Oxford University Press, 1995), 276–299; Genter, *Late Modernism*, 93–100.

94. See Genter, *Late Modernism*.

95. W. M. McClay, *The Masterless: Self and Society in Modern America* (Chapel Hill: University of North Carolina Press, 1994), 189–225. Key texts to interrogate this subject at the time were M. Horkheimer and T. Adorno, *Dialectic of Enlightenment* (1944; repr., New York: Herder & Herder, 1972); and H. Arendt, *The Origins of Totalitarianism* (1951; repr., New York: Harcourt, Brace, 1966).

96. Genter, *Late Modernism*, 96. For more on the relocation of the Frankfurt School to America, see T. Wheatland, *The Frankfurt School in Exile* (Minneapolis: University of Minnesota Press, 2009); M. Jay, *The Dialectical Imagination: A History of the Frankfurt School and the Institute of Social Research, 1923–1950* (Boston: Little, Brown, 1973).

97. E. Fromm, *The Fear of Freedom* (London: Routledge & Kegan Paul, 1942), 218. For more on Fromm's life, see L. J. Friedman and A. M. Schreiber, *The Lives of Erich Fromm: Love's Prophet* (New York: Columbia University Press, 2013).

98. Fromm, *Fear of Freedom*, 239.

99. Rainer Funk discusses Fromm's personal involvement with religion. See R. Funk, *Erich Fromm, His Life and Ideas: An Illustrated Biography* (New York: Continuum, 2000).

100. Fromm, *Fear of Freedom*, 243.

101. Fromm, *Fear of Freedom*, 80.

102. Fromm, *Fear of Freedom*, 87.

103. Fromm, *Fear of Freedom*, 87–88.

104. Fromm, *Fear of Freedom*, 226–227.

105. P. Goodman, "The Political Meaning of Some Recent Revisions of Freud," *politics* 2, no. 7 (July 1945): 197–203; C. Wright Mills and P. J. Salter, "The Barricade and the Bedroom," *politics* 2, no. 10 (October 1945): 313–315.

106. Goodman, "Political Meaning," 197. See also P. Goodman, "Revolution, Sociolatry and War," *politics* 2, no. 12 (December 1945): 378.

107. Goodman, "Revolution, Sociolatry and War," 379, 380.

108. Goodman, "Political Meaning," 202.

109. See Turner, *Adventures in the Orgasmatron*, 244–251.

110. Mills and Salter, "Barricade and the Bedroom," 314. For more on this debate, see Byers, *Charles Olson and American Modernism*, 45–63.

111. Mills and Salter, "Barricade and the Bedroom," 314.

112. Mills and Salter, "Barricade and the Bedroom," 314.

113. Mills and Salter, "Barricade and the Bedroom," 315.

114. Mills and Salter, "Barricade and the Bedroom," 315.

115. R. V. Chase, "The Huxley-Heard Paradise," *Partisan Review* 10, no. 2 (1943): 155.

116. Chase, "Huxley-Heard Paradise," 153.

117. Chase, "Huxley-Heard Paradise," 155.
118. Chase, "Huxley-Heard Paradise," 154.
119. Chase, "Huxley-Heard Paradise," 154.
120. Chase, "Huxley-Heard Paradise," 152.
121. See D. Bradshaw, "Huxley's Slump: Planning, Eugenics, and the 'Ultimate Need' of Stability," in *The Art of Literary Biography*, ed. John Batchelor (Oxford: Clarendon, 1995), 151–172; D. Bradshaw, *The Hidden Huxley: Contempt and Compassion for the Masses, 1920–36* (London: Faber & Faber, 1994), xvii. For more on Huxley's interest in planning, see J. Woiak, "Designing a Brave New World: Eugenics, Politics, and Fiction," *Public Historian* 29, no. 3 (August 1, 2007): 105–129.
122. See R. S. Deese, *We Are Amphibians: Julian and Aldous Huxley on the Future of Our Species* (Oakland: University of California Press, 2015), 92.
123. A. Huxley, *Ends and Means: An Inquiry into the Nature of Ideals and into Methods Employed for Their Realization* (New York: Harper & Brothers, 1937), 3–4.
124. Huxley, *Ends and Means*, 23.
125. Huxley, *Ends and Means*, 311.
126. J. D. Unwin, *Sex and Culture* (London: Oxford University Press, 1934).
127. Huxley, *Ends and Means*, 311.
128. Huxley, *Ends and Means*, 311.
129. Huxley, *Ends and Means*, 315.
130. Huxley, *Ends and Means*, 316.
131. Unwin, *Sex and Culture*, xv.
132. Huxley, *Ends and Means*, 314.
133. Huxley, *Ends and Means*, 315.
134. Huxley, *Ends and Means*, 317.
135. Huxley, *Ends and Means*, 319.
136. A. Huxley, introduction to *Hopousia; or, The Sexual and Economic Foundations of a New Society*, by J. D. Unwin (London: George Allen & Unwin, 1940), 13–29.
137. See J. D. Unwin, *Hopousia; or, The Sexual and Economic Foundations of a New Society* (London: George Allen & Unwin, 1940).
138. See Huxley, introduction to *Hopousia*, 26.
139. Huxley, introduction to *Hopousia*, 28.
140. Huxley, introduction to *Hopousia*, 25.
141. For an extended comparison between Julian Huxley and Aldous Huxley, see Deese, *We Are Amphibians*. See also P. Anker, *Imperial Ecology: Environmental Order in the British Empire, 1895–1945* (Cambridge, MA: Harvard University Press, 2001), 114.
142. J. Huxley, *A Scientist among the Soviets* (London: Harper & Brothers, 1932).
143. See J. Huxley, *TVA, Adventure in Planning* (Surrey: Architectural Press, 1943).
144. Mumford, *Conduct of Life*, 240. See also L. S. Wittner, *The Struggle against the Bomb*, vol. 1, *One World or None: A History of the World Nuclear Disarmament Movement through 1953* (Stanford, CA: Stanford University Press, 1993), 66–71. For Mumford's response to the threat of the atomic bomb, see L. Mumford, "Gentlemen: You Are Mad!," *Saturday Review of Literature* 29, no. 9 (March 2, 1946): 5–6.
145. G. Sluga, "UNESCO and the (One) World of Julian Huxley," *Journal of World History* 21, no. 3 (September 1, 2010): 396. For more on the idea of the "World State," see G. Sluga, *Internationalism in the Age of Nationalism* (Philadelphia: University of

Pennsylvania Press, 2013); G. Sluga, *The Nation, Psychology, and International Politics, 1870–1919* (New York: Palgrave Macmillan, 2006); J. A. Pemberton, *Global Metaphors: Modernity and the Quest for One World* (London: Pluto, 2001); J. A. Pemberton, "New Worlds for Old: The League of Nations in the Age of Electricity," *Review of International Studies* 28, no. 2 (April 1, 2002): 311–336; A. Bashford, *Global Population: History, Geopolitics, and Life on Earth* (New York: Columbia University Press, 2013); J. Partington, *Building Cosmopolis: The Political Thought of H. G. Wells* (Aldershot: Ashgate, 2003). Glenda Sluga points out how Huxley understood UNESCO as an extension of the British Empire. See Sluga, "UNESCO and the (One) World," 393–418.

146. Huxley's publishing record was extensive. For a detailed bibliography, see J. R. Baker and J. P. Green, *Julian Huxley, Scientist and World Citizen, 1887 to 1975: A Biographical Memoir* (Paris: UNESCO, 1978).

147. J. Huxley, *Religion without Revelation* (New York: Harper & Brothers, 1957), 85.

148. For more on Huxley, see M. Swetlitz, "Julian Huxley and the End of Evolution," *Journal of the History of Biology* 28, no. 2 (July 1, 1995): 181–217; R. M. Gascoigne, "Julian Huxley and Biological Progress," *Journal of the History of Biology* 24, no. 3 (October 1, 1991): 433–455.

149. H. G. Wells, J. Huxley, and G. P. Wells, *The Science of Life* (1931; repr., London: Casell, 1946), 4.

150. Wells, Huxley, and Wells, *Science of Life*, 6.

151. Wells, Huxley, and Wells, *Science of Life*, 1016.

152. Wells, Huxley, and Wells, *Science of Life*, 1015.

153. Huxley, *Religion without Revelation*, 91.

154. Huxley, *Religion without Revelation*, 91.

155. Huxley, *Religion without Revelation*, 56.

156. Huxley, *Religion without Revelation*, 220.

157. Huxley, *Religion without Revelation*, 220–221.

158. Huxley, *Religion without Revelation*, 221.

159. J. Huxley, *If I Were Dictator* (New York: Harper & Brothers, 1934), 116.

160. Huxley, *If I Were Dictator*, 18.

161. Huxley, *If I Were Dictator*, 24.

162. J. Huxley, "On Living in a Revolution," *Harper's Monthly Magazine* 185, no. 1108 (1942): 344–345. This was republished as an essay in J. Huxley, *On Living in a Revolution* (New York: Harper & Brothers, 1944).

163. Huxley, "On Living in a Revolution," 344–345.

164. Julian Huxley described the yardstick metaphor as a distinctly American expression during a panel discussion on the TVA in 1937. C. Close, J. S. Huxley, and G. B. Barbour, "The Tennessee Valley Project: Discussion," *Geographical Journal* 89, no. 5 (May 1, 1937): 405–408.

165. Huxley, "On Living in a Revolution," 344–345.

166. Huxley, *Evolutionary Ethics*, 55.

167. Huxley, *Evolutionary Ethics*, 64.

168. See Sluga, "UNESCO and the (One) World," 393–418.

169. For more on how Einstein's and Bergson's philosophies informed their political ideas, see J. Canales, "Einstein, Bergson, and the Experiment That Failed: Intellectual Cooperation at the League of Nations," *MLN* 120, no. 5 (2005): 1168–1191.

170. V. Vernadsky, *The Biosphere*, ed. M. K. A. McMenamin (New York: Copernicus, 1998). Alison Bashford describes how Vernadsky's notion of the biosphere initiated visions of an integrated world. See Bashford, *Global Population*, 170.

171. For more on Wells's internationalism, see J. S. Partington, *Building Cosmopolis: The Political Thought of H. G. Wells* (Aldershot: Ashgate, 2003).

172. H. G. Wells, *A Modern Utopia*, ed. G. Claeys and P. Parrinder (1905; repr., London: Penguin Books, 2005), 56.

173. Wells, Huxley, and Wells, *Science of Life*, 1016, quoted in Anker, *Imperial Ecology*, 114.

174. For a discussion of Wells's internationalism in relation to the new field of ecology, see Bashford, *Global Population*, 175–179.

175. J. Huxley, *UNESCO: Its Purpose and Its Philosophy* (Washington, DC: Public Affairs Press, 1947).

176. Huxley, *UNESCO*, 17.

177. J. Huxley, "Education as a Social Function," in *On Living in a Revolution* (New York: Harper, 1944), 192, 193.

178. Huxley, *UNESCO*, 61.

179. Huxley, *UNESCO*, 13.

180. J. Huxley, introduction to *The Phenomenon of Man*, by P. Teilhard de Chardin (New York: Harper, 1959), 11.

181. Huxley, introduction to *Phenomenon of Man*, 11.

182. For a history of the development of the term *noosphere*, see P. R. Samson and D. Pitt, *The Biosphere and Noosphere Reader Global Environment, Society and Change* (London: Routledge, 2002).

183. Within *The Phenomenon of Man*, Teilhard de Chardin posited two forms of energy: physical energy, which was measurable and calculable, and "psychic energy," which increased with complexity. Huxley, introduction to *Phenomenon of Man*, 17.

184. Huxley, introduction to *Phenomenon of Man*, 12.

185. Huxley, introduction to *Phenomenon of Man*, 17.

186. Sluga, "UNESCO and the (One) World," 408.

187. See J. Huxley, "UNESCO: Its Purpose and Philosophy," *Free World* 7, no. 5 (December 1946): 27–30; F. Meyer, "Long-Term Hope for Humanity: UNESCO: Its Purpose and Its Philosophy," *New York Times*, September 7, 1947, 32.

188. See J. Stevens, *Storming Heaven: LSD and the American Dream* (New York: Atlantic Monthly Press, 1987); M. A. Lee and B. Shlain, *Acid Dreams: The Complete Social History of LSD: The CIA, the Sixties, and Beyond* (New York: Grove Weidenfeld, 1992).

189. A. Huxley, *The Doors of Perception* (New York: Harper & Brothers, 1954).

190. See J. Huxley, "The Crisis in Man's Destiny," *Playboy* 14 (January 1967): 93–94, 212–217.

191. Huxley, "Crisis in Man's Destiny," 213.

192. Huxley, "Crisis in Man's Destiny," 213.

Epilogue

1. C. Olson, *The Maximus Poems*, ed. G. F. Butterick (Berkeley: University of California Press, 1983), 461.

2. J. Carter, "Report to the American People on Energy" (February 2, 1977), in *Public Papers of the Presidents of the United States, Jimmy Carter, 1977*, bk. 1, *January 20 to June 24, 1977* (Washington, DC: US Government Printing Office, 1977), 69–77.

3. "What Price Energy?," *Newsweek*, May 2, 1977, 12, quoted in B. J. Schulman, *The Seventies: The Great Shift in American Culture, Society and Politics* (New York: Free Press, 2001), 127.

4. J. Carter, "Energy and National Goals" (July 15, 1979), in *Public Papers of the Presidents of the United States, Jimmy Carter, 1979*, bk. 2, *June 23 to December 31, 1979* (Washington, DC: US Government Printing Office, 1979), 1235.

5. Carter, "Energy and National Goals," 1237.

6. Daniel Horowitz has pointed to the importance of these thinkers in influencing Carter's "Crisis of Confidence" speech. See D. Horowitz, *The Anxieties of Affluence: Critiques of American Consumer Culture, 1939–1979* (Amherst: University of Massachusetts Press, 2004), 203–245. For other interpretations of Carter's "Crisis of Confidence" speech, see L. Ribuffo, "'Malaise' Revisited: Jimmy Carter and the Crisis of Confidence," in *The Liberal Persuasion: Arthur Schlesinger, Jr., and the Challenge of the American Past*, ed. J. P. Diggins (Princeton, NJ: Princeton University Press, 1997), 164–184; K. E. Morris, *Jimmy Carter, American Moralist* (Athens: University of Georgia Press, 1996), 251–262.

7. Carter, "Energy and National Goals," 1238–1239.

8. See, e.g., M. J. Graetz, *The End of Energy: The Unmaking of America's Environment, Security, and Independence* (Cambridge, MA: MIT Press, 2011).

9. "Nation's Newspaper Editorialists Split over Carter's Energy Talk," *New York Times*, July 17, 1979, A13.

10. See J. Carter, "Address to the Nation on Energy," April 18, 1977, American Presidency Project, https://www.presidency.ucsb.edu/documents/address-the-nation-energy.

11. N. Wiener, *Cybernetics, or Control and Communication in the Animal and the Machine* (1948; repr., Cambridge, MA: MIT Press 1965), 132.

12. K. Hayles, *How We Became Posthuman: Virtual Bodies in Cybernetics, Literature, and Informatics* (Chicago: University of Chicago Press, 1999).

13. C. N. Daggett, *The Birth of Energy: Fossil Fuels, Thermodynamics, and the Politics of Work* (Durham, NC: Duke University Press, 2019), 190.

14. N. Labanca, "Ontological Fallacies Linked to Energy, Information and Related Technologies," in *Complex Systems and Social Practices in Energy Transitions: Framing Energy Sustainability in the Time of Renewables*, ed. N. Labanca (Cham, Switzerland: Springer International, 2017), 171–206.

15. J. Rinkinen, E. Shove, and J. Torriti, *Energy Fables: Challenging the Ideas of the Energy Sector* (London: Routledge, 2019), 2.

16. Daggett, *Birth of Energy*, 190.

BIBLIOGRAPHY

Archives

Chase, Stuart. Papers. Manuscript Division, Library of Congress, Washington, DC.

Dreiser, Theodore. Papers. Annenberg Rare Book and Manuscript Library, University of Pennsylvania, Philadelphia, Pennsylvania.

Loeb, Jacques. Papers. Manuscript Division, Library of Congress, Washington, DC.

Mumford, Lewis. Papers. Annenberg Rare Book and Manuscript Library, University of Pennsylvania, Philadelphia, Pennsylvania.

Osborn, Henry Fairfield. Papers. Research Library, American Museum of Natural History, New York, New York.

The Youth Movement Archive, London School of Economics, London, England.

Sources

Acker, Antoine. "A Different Story of the Anthropocene: Brazil's Post-Colonial Quest for Oil (1930–1975)." *Past and Present* 249, no. 1 (2020): 167–211.

Ackermann, Marsha E. *Cool Comfort: America's Romance with Air-Conditioning.* Washington, DC: Smithsonian Institution Press, 2002.

Adams, Brooks. "The Heritage of Henry Adams" (1919). In Henry Adams, *The Degradation of the Democratic Dogma*, 1–122. New York: Harper & Row, 1969.

Adams, Brooks. *The Law of Civilization and Decay: An Essay on History.* With an introduction by Charles A. Beard. 1895. Reprint, New York: Alfred A. Knopf, 1943.

Adams, Brooks. *The Theory of Social Revolutions.* New York: Macmillan, 1913.

Adams, Henry. *The Education of Henry Adams.* Edited by Ira B. Nadel. 1907. Reprint, Oxford: Oxford University Press, 1999.

Adams, Henry. *A Letter to American Teachers of History.* Washington, DC: J. H. Furst, 1910.

Adams, Henry. *Mont-Saint-Michel and Chartres* (1904). In *Henry Adams: Novels, Mont Saint Michel, The Education*, edited by Ernest Samuels and Jayne N. Samuels, 337–714. New York: Literary Classics of the United States, 1983.

Akin, William E. *Technocracy and the American Dream: The Technocrat Movement, 1900–1941.* Berkeley: University of California Press, 1977.

Alexander, Jennifer Karns. *The Mantra of Efficiency: From Waterwheel to Social Control.* Baltimore: Johns Hopkins University Press, 2008.

Allen, Garland E. "T. H. Morgan and the Emergence of a New American Biology." *Quarterly Review of Biology* 44, no. 2 (June 1, 1969): 168–188.

Alpers, Benjamin Leontief. *Dictators, Democracy, and American Public Culture: Envisioning the Totalitarian Enemy, 1920–1950s.* Chapel Hill: University of North Carolina Press, 2003.

Anderson, Larry. *Benton MacKaye: Conservationist, Planner, and Creator of the Appalachian Trail.* Baltimore: Johns Hopkins University Press, 2002.

Anderson, Thornton. *Brooks Adams, Constructive Conservative.* Ithaca, NY: Cornell University Press, 1951.

Anker, Peder. "Buckminster Fuller as Captain of Spaceship Earth." *Minerva* 45, no. 4 (2007): 417–434.

Anker, Peder. *From Bauhaus to Ecohouse: A History of Ecological Design.* Baton Rouge: Louisiana State University Press, 2010.

Anker, Peder. *Imperial Ecology: Environmental Order in the British Empire, 1895–1945.* Cambridge, MA: Harvard University Press, 2001.

Ardzrooni, Leon. "Veblen and Technocracy." *Living Age* 344 (March 1933): 29–40.

Arendt, Hannah. *The Origins of Totalitarianism.* 1951. Reprint, New York: Harcourt, Brace, 1966.

Armstrong, Tim. *Modernism, Technology, and the Body: A Cultural Study.* Cambridge: Cambridge University Press, 1998.

Armstrong, Tim. "Social Credit Modernism." *Critical Quarterly* 55, no. 2 (2013): 50–65.

"Atmosphere and Sun Are Linked as Static Cause." *Brooklyn Eagle*, April 16, 1927, C7.

Ayres, P. G. *Shaping Ecology: The Life of Arthur Tansley.* Chichester: Wiley-Blackwell, 2012.

Baird, A. S. "Wonderful Findings of College Professors." *Nashville Banner*, Tennessee, January 18, 1913, 11.

Baker, John Randal, and Jens-Peter Green. *Julian Huxley, Scientist and World Citizen, 1887 to 1975: A Biographical Memoir.* Paris: UNESCO, 1978.

Banks, Erik C. *Ernst Mach's World Elements: A Study in Natural Philosophy.* Dordrecht: Kluwer Academic, 2003.

Banta, Martha. *Taylored Lives: Narrative Productions in the Age of Taylor, Veblen, and Ford.* Chicago: University of Chicago Press, 1993.

Baratta, Joseph Preston. *The Politics of World Federation.* Westport, CT: Praeger, 2004.

Barrett, Ross, and Daniel Worden. *Oil Culture.* Minneapolis: University of Minnesota Press, 2014.

Barrett, Ross, and Daniel Worden, eds. "Oil Cultures." Special issue, *Journal of American Studies* 46, no. 2 (2012).

Bashford, Alison. *Global Population: History, Geopolitics, and Life on Earth.* New York: Columbia University Press, 2014.

Beard, George Miller. *American Nervousness: Its Causes and Consequences, A Supplement to Nervous Exhaustion (Neurasthenia).* New York: G. P. Putnam's Sons, 1881.

Beasley, Rebecca. *Ezra Pound and the Visual Culture of Modernism.* Cambridge: Cambridge University Press, 2007.

Belgrad, Daniel. *The Culture of Spontaneity: Improvisation and the Arts in Postwar America*. Chicago: University of Chicago Press, 1999.

Bergson, Henri. *The Meaning of the War: Life and Matter in Conflict*. London: T. Fisher Unwin, 1915.

Beringause, Arthur. *Brooks Adams: A Biography*. New York: Alfred A. Knopf, 1955.

Berthkinert, A. "New Science Suggested by Harvard Man, Cosmecology: Would Bind Together Sciences Governing Earth." *Waycross Journal Herald*, June 11, 1934, 3.

Black, Brian. "Oil for Living: Petroleum and American Conspicuous Consumption." *Journal of American History* 99, no. 1 (2012): 40–50.

Black, Brian. *Petrolia: The Landscape of America's First Oil Boom*. Baltimore: Johns Hopkins University Press, 2000.

Blake, Casey Nelson. *Beloved Community: The Cultural Criticism of Randolph Bourne, Van Wyck Brooks, Waldo Frank and Lewis Mumford*. Chapel Hill: University of North Carolina Press, 1990.

Blakeslee, Alton L. "Warm Earth Cycle in West Might Make Russia Kingpin." *Eugene Register Guard*, November 15, 1949, 6.

Blakeslee, Alton L. "'Warming Up' of Earth May Affect This Country." *Ottawa Citizen*, November 1, 1949, 17.

Bourne, Randolph. "A Moral Equivalent for Universal Military Service." *New Republic* 7, no. 87 (July 1, 1916): 217–219.

Bourne, Randolph. "Twilight of Idols." *Seven Arts* 11 (October 1917): 688–702.

Boyer, Paul S. *By the Bomb's Early Light: American Thought and Culture at the Dawn of the Atomic Age*. New York: Pantheon, 1985.

Boyer, Paul S. *Urban Masses and Moral Order in America*. Cambridge, MA: Harvard University Press, 1975.

Bradshaw, David. *The Hidden Huxley: Contempt and Compassion for the Masses, 1920–36*. London: Faber & Faber, 1994.

Bradshaw, David. "Huxley's Slump: Planning, Eugenics, and the 'Ultimate Need' of Stability." In *The Art of Literary Biography*, edited by John Batchelor, 151–172. Oxford: Clarendon, 1995.

Bramwell, Anna. *Ecology in the 20th Century: A History*. New Haven, CT: Yale University Press, 1989.

Brick, Howard. *Transcending Capitalism: Visions of a New Society in Modern American Thought*. Ithaca, NY: Cornell University Press, 2006.

Briggs, Asa. *The Birth of Broadcasting*. Oxford: Oxford University Press, 1961.

Brinkley, Alan. *Voices of Protest: Huey Long, Father Coughlin, and the Great Depression*. New York: Knopf, 1982.

Bruère, Robert. "Giant Power—Region Builder." *Survey Graphic* 54, no. 4 (1925): 161–164.

Bruère, Robert. "Pandora's Box." *Survey Graphic* 51, no. 11 (March 1, 1924): 557–560.

Buell, Frederick. "A Short History of Oil Cultures: Or, the Marriage of Catastrophe and Exuberance." *Journal of American Studies* 46, no. 2 (2012): 273–293.

Burich, Keith R. "Henry Adams, the Second Law of Thermodynamics, and the Course of History." *Journal of the History of Ideas* 48, no. 3 (July 1, 1987): 467–482.

Burkitt, Brian, and Frances Hutchinson. *The Political Economy of Social Credit and Guild Socialism.* London: Routledge, 1997.

Burnham, John C. "From Avant-Garde to Specialism: Psychoanalysis in America." *Journal of the History of Behavioural Sciences* 15 (1979): 128–134.

Burns, Jennifer. *Goddess of the Market: Ayn Rand and the American Right.* Oxford: Oxford University Press, 2009.

Burroughs, John. *Camping and Tramping with President Roosevelt.* Carlisle, MA: Applewood Books, 1907.

Burwick, Frederick, and Paul Douglass. *The Crisis in Modernism: Bergson and the Vitalist Controversy.* Cambridge: Cambridge University Press, 1992.

Byers, Mark. *Charles Olson and American Modernism: The Practice of the Self.* Oxford: Oxford University Press, 2018.

Cain, Joe. "Julian Huxley, General Biology and the London Zoo, 1935–42." *Notes and Records of the Royal Society* 64, no. 4 (December 20, 2010): 359–378.

Cameron, Laura, and John Forrester. "'A Nice Type of English Scientist': Tansley and Freud." *History Workshop Journal* 48 (1999): 65–100.

Campbell, Joseph. *The Hero with a Thousand Faces.* 1949. Reprint, Princeton, NJ: Princeton University Press, 1972.

Canales, Jimena. "Einstein, Bergson, and the Experiment That Failed: Intellectual Cooperation at the League of Nations." *MLN* 120, no. 5 (2005): 1168–1191.

Carlowicz, Michael J., and Ramon E. Lopez. *Storms from the Sun: The Emerging Science of Space Weather.* Washington, DC: Joseph Henry, 2002.

Carrier, Louis. "Find: Hidden Forces Run Animal Cycles." *New York Times*, August 16, 1931, 32.

Carrier, Louis. "Fluctuations in Animal Life Studied for Effects on Man: Cyclical Periods of Abundance and Scarcity Arouse Deep Speculations among Natural Scientists Meeting in Canada." *New York Times*, August 30, 1931, xx4.

Carson, Mina J. "The Evolution of Brooks Adams." *Biography* 6, no. 2 (1983): 95–116.

Carter, Jimmy. "Energy and National Goals" (July 15, 1979). In *Public Papers of the Presidents of the United States, Jimmy Carter, 1979,* book 2, *June 23–December 31, 1979,* 1235–1241. Washington, DC: US Government Printing Office, 1979.

Carter, Jimmy. "Report to the American People on Energy" (February 2, 1977). In *Public Papers of the Presidents of the United States, Jimmy Carter, 1977,* book 1, *January 20 to June 24, 1977,* 69–77. Washington, DC: US Government Printing Office, 1977.

Carter, Simon. *Rise and Shine: Sunlight, Technology, and Health.* Oxford: Berg, 2007.

Carus, Paul. "A Monistic Conception of Consciousness: In Reply to Mr. Ayton Wilkinson's Article on 'Will-Force' and Mr. Montague's 'Are Mental Processes in Space?'" *Monist* 18, no. 1 (1908): 30–45.

Cassirer, Ernst. *Substance and Function and Einstein's Theory of Relativity.* New York: Dover, 1953.

"The Champ." *Time* 66, no. 25 (December 19, 1955): 66.

Chappell, John E. "Huntington and His Critics: The Influence of Climate on Civilization." PhD diss., University of Kansas, 1976.

Chase, Richard V. "The Huxley-Heard Paradise." *Partisan Review* 10, no. 2 (1943): 143–158.

Chase, Stuart. "Coals to Newcastle." *Survey Graphic* 24, no. 4 (May 1, 1925): 143–146.

Chase, Stuart. *The Economy of Abundance*. New York: Macmillan, 1934.
Chase, Stuart. *A New Deal*. New York: Macmillan, 1932.
Chase, Stuart. "The New Energy." *Nation* 161, no. 25 (December 22, 1945): 709–710.
Chase, Stuart. *One Billion Wild Horses*. New York: League for Economic Democracy, 1930.
Chase, Stuart. "Planning for Natural Resources in America." In *Third World Power Conference Transactions*, 1:451–483. Washington, DC: US Government Printing Office, 1936.
Chase, Stuart. "Portrait of a Radical." *Century Magazine* 108, no. 3 (1924): 295–301.
Chase, Stuart. *Technocracy: An Interpretation*. New York: John Day Pamphlets, 1933.
Chase, Stuart. *The Tragedy of Waste*. New York: Macmillan, 1925.
Chase, Stuart, and Marian Tyler Chase. *Mexico: A Study of Two Americas*. London: John Lane the Bodley Head, 1932.
Chatterjee, Elizabeth. "The Asian Anthropocene: Electricity and Fossil Developmentalism." *Journal of Asian Studies* 79, no. 1 (2020): 3–24.
Chitty, Dennis. *Do Lemmings Commit Suicide? Beautiful Hypotheses and Ugly Facts*. Oxford: Oxford University Press, 1996.
Choi, Tina Young. "Forms of Closure: The First Law of Thermodynamics and Victorian Narrative." *ELH* 74 (2007): 301–322.
Christie, Jean. "Giant Power: A Progressive Proposal of the Nineteen-Twenties." *Pennsylvania Magazine of History and Biography* 96, no. 4 (1972): 480–507.
Ciepley, David. *Liberalism in the Shadow of Totalitarianism*. Cambridge, MA: Harvard University Press, 2006.
Clark, Tom. *Charles Olson: The Allegory of a Poet's Life*. New York: Norton, 1991.
Clarke, Bruce. *Energy Forms: Allegory and Science in the Era of Classical Thermodynamics*. Ann Arbor: University of Michigan Press, 2001.
Clarke, Bruce, and Linda Dalrymple Henderson. *From Energy to Information: Representation in Science and Technology, Art, and Literature*. Stanford, CA: Stanford University Press, 2002.
Close, Charles, Julian S. Huxley, and George B. Barbour. "The Tennessee Valley Project: Discussion." *Geographical Journal* 89, no. 5 (May 1, 1937): 405–408.
"Coal and Morals." *Literary Digest* 61, no. 12 (June 21, 1919): 23–24.
Conference on Conservation of Natural Resources, W. J. McGee, and Newton Crain Blanchard. *Proceedings of a Conference of Governors in the White House, Washington, D.C., May 13–15, 1908*. Washington, DC: US Government Printing Office, 1909.
Conference on Cycles. *Reports of the Conferences on Cycles*. Washington, DC: Carnegie Institution, 1929.
Conn, Steven. *Americans against the City: Anti-urbanism in the Twentieth Century*. Oxford: Oxford University Press, 2014.
Cooke, G. W. "Science and Socialism." *Wall Street Journal*, September 25, 1912.
Coopersmith, Jonathan. *The Electrification of Russia, 1880–1926*. Ithaca, NY: Cornell University Press, 1992.
Coopersmith, Jonathan. "Technology Transfer in Russian Electrification, 1870–1925." *History of Technology* 13 (1991): 214–234.
Cooter, Roger, and Stephen Pumfrey. "Separate Spheres and Public Places: Reflections on the History of Science Popularization and Science in Public Culture." *History of Science* 32, no. 97 (1994): 237–267.

Cotkin, George. *William James, Public Philosopher.* Baltimore: Johns Hopkins University Press, 1990.

Coughlin, Charles E. "The God of Gold." In *Father Coughlin's Radio Discourses, 1931–32,* 161–173. Royal Oak, MI: Radio League of the Little Flower, 1932.

Coughlin, Charles E. *Money! Questions and Answers.* Royal Oak, MI: National Union for Social Justice, 1936.

Cousins, Norman. "Modern Man Is Obsolete." *Saturday Review* 28, no. 32 (August 18, 1945): 5–9.

Cowan, Michael J. *Cult of the Will: Nervousness and German Modernity.* University Park: Pennsylvania State University Press, 2008.

Cox, Stephen D. *The Woman and the Dynamo: Isabel Paterson and the Idea of America.* New Brunswick, NJ: Transaction, 2004.

Creese, Walter L. *TVA's Public Planning: The Vision, the Reality.* Knoxville: University of Tennessee Press, 1990.

Crick, Walter, and Frederick Soddy. *Abolish Private Money, or Drown in Debt.* London: Sault, 1939.

"Crime Is Laid to Energies 'Misdirected': Poverty and Illiteracy Not Biggest Factors, Says Hoover Aide." *Washington Post,* May 14, 1938, X17.

Crosby, Alfred. *Children of the Sun: A History of Humanity's Unappeasable Appetite for Energy.* New York: W. W. Norton, 2006.

Crowcroft, Peter. *Elton's Ecologists: A History of the Bureau of Animal Population.* Chicago: University of Chicago Press, 1991.

Crutzen, Paul J., and Eugene F Stoermer. "The Anthropocene." *Global Change Newsletter* 41 (2000): 17–18.

Daggett, Cara New. *The Birth of Energy: Fossil Fuels, Thermodynamics, and the Politics of Work.* Durham, NC: Duke University Press, 2019.

Dalton, Kathleen. *Theodore Roosevelt: A Strenuous Life.* New York: Alfred A. Knopf, 2002.

Dalton, Kathleen. "Why America Loved Teddy Roosevelt: Or, Charisma Is in the Eyes of the Beholders." In *Our Selves/Our Past: Psychological Approaches to American History,* edited by Robert J. Brugger, 269–291. Baltimore: Johns Hopkins University Press, 1981.

Dalton, Thomas Carlyle. *Becoming John Dewey: Dilemmas of a Philosopher and Naturalist.* Bloomington: Indiana University Press, 2002.

Daston, Lorraine. "The Theory of Will versus the Science of Mind." In *The Problematic Science: Psychology in Nineteenth-Century Thought,* edited by William R. Woodward and Timothy G. Ash, 88–115. New York: Praeger, 1982.

Davis, W. "Science Sees Improvement in 1931 Radio Reception." *Pittsburgh Press,* December 30, 1930, 21.

Deese, R. S. *We Are Amphibians: Julian and Aldous Huxley on the Future of Our Species.* Oakland: University of California Press, 2015.

Deltete, Robert. "Helm and Boltzmann: Energetics at the Lübeck *Naturforscherversammlung.*" *Synthese* 119, no. 1/2 (January 1, 1999): 45–68.

"Democracy in the Fining-Pot." *Literary Digest* 51, no. 7 (August 15, 1915): 298–299.

Denton, Sally. *The Plots against the President: FDR, a Nation in Crisis, and the Rise of the American Right.* New York: Bloomsbury, 2012.

Dewey, Edward R. *Cycles: The Mysterious Forces That Trigger Events.* New York: Hawthorn Books, 1971.
Dewey, John. *Art as Experience.* New York: Minton, Balch, 1934.
Dewey, John. "Force and Coercion." *International Journal of Ethics* 26, no. 3 (April 1, 1916): 359–367.
Dewey, John. "Force, Violence and Law." *New Republic* 5, no. 64 (January 22, 1916): 295–297.
Dewey, John. *Impressions of Soviet Russia and the Revolutionary World: Mexico—China—Turkey.* 1929. Reprint, New York: Teachers College, Columbia University, 1956.
Dewey, John. "On Understanding the Mind of Germany." *Atlantic Monthly* 117 (February 1916): 251–262.
Diggins, John P. *The Promise of Pragmatism: Modernism and the Crisis of Knowledge and Authority.* Chicago: University of Chicago Press, 1994.
Director, Aaron. *The Economics of Technocracy.* Edited by Harry D. Gideonse. Chicago: University of Chicago Press, 1933.
Donaldson, Scott, and R. H. Winnick. *Archibald MacLeish: An American Life.* Boston: Houghton Mifflin, 1992.
Donovan, Timothy Paul. *Henry Adams and Brooks Adams: The Education of Two American Historians.* Norman: University of Oklahoma Press, 1961.
Dooley, Patrick K. "Public Policy and Philosophical Critique: The William James and Theodore Roosevelt Dialogue on Strenuousness." *Transactions of the Charles S. Peirce Society* 37, no. 2 (April 1, 2001): 161–177.
Dorfman, Joseph. *Thorstein Veblen and His America.* New York: Viking, 1934.
Dorman, Robert L. *Revolt of the Provinces: The Regionalist Movement in America, 1920–1945.* Chapel Hill: University of North Carolina Press, 1993.
Dorsey, Leroy G. *We Are All Americans, Pure and Simple: Theodore Roosevelt and the Myth of Americanism.* Tuscaloosa: University of Alabama Press, 2007.
Douglas, Susan J. *Listening In: Radio and the American Imagination.* New York: Random House, 1999.
Douglass, A. E. "Evidence of Climatic Effects in the Annual Rings of Trees." *Ecology* 1, no. 1 (1920): 24–32.
Drakeford, Mark. *Social Movements and Their Supporters: The Green Shirts in England.* New York: St. Martin's, 1997.
Duffus, Robert L. "Jacques Loeb: Mechanist." *Century Magazine* 108, no. 3 (July 1924): 374–383.
"Ebbing Tide." *Time* 36, no. 20 (November 11, 1940): 43.
Eddington, Arthur Stanley. *New Pathways in Science.* Cambridge: Cambridge University Press, 1935.
Edge, Charles N. "The Measurements of Technology." *Living Age,* January 1933, 399–408.
Eichengreen, Barry J. *Golden Fetters: The Gold Standard and the Great Depression, 1919–1939.* Oxford: Oxford University Press, 1992.
Ekbladh, David. "Meeting the Challenge from Totalitarianism: The Tennessee Valley Authority as a Global Model for Liberal Development, 1933–1945." *International History Review* 32, no. 1 (2010): 47–67.

Elton, Charles S. *Animal Ecology*. 1927. Reprint, Chicago: University of Chicago Press, 2001.
Elton, Charles S. "Periodic Fluctuations in the Numbers of Animals: Their Causes and Effects." *Journal of Experimental Biology* 2 (1924): 119–163.
"Energy of the Human Will." *Literary Digest* 36, no. 360 (February 15, 1908): 223.
Engerman, David C. "John Dewey and the Soviet Union: Pragmatism Meets Revolution." *Modern Intellectual History* 3, no. 1 (2006): 33–63.
Engerman, David C. *Modernization from the Other Shore: American Intellectuals and the Romance of Russian Development*. Cambridge, MA: Harvard University Press, 2003.
Erskine, John. "Erginette: A Future Helen amid the Topless Towers of Technocracy?" *Collier's Weekly*, March 25, 1933, 10–11, 28–30.
"Ethics for an Atomic Age." *Saturday Review* 28, no. 35 (September 1, 1945): 18.
F., A. L. "A Letter to a German-American Friend." *Outlook*, January 6, 1915, 13–16.
Fangerau, Heiner. "Biology and War—American Biology and International Science." *History and Philosophy of the Life Sciences* 29, no. 4 (2007): 395–427.
Fangerau, Heiner. "From Mephistopheles to Isaiah: Jacques Loeb, Technical Biology and War." *Social Studies of Science* 39, no. 2 (2009): 229–256.
Fangerau, Heiner. "Monism, Racial Hygiene, and National Socialism." In *Monism, Science, Philosophy, Religion, and the History of a Worldview*, edited by Todd H. Weir, 223–249. New York: Palgrave Macmillan, 2012.
Farca, Paula Anca, ed. *Energy in Literature: Essays on Energy and Its Social and Environmental Implications in Twentieth and Twenty-First Century Literary Texts*. Oxford: TrueHeart, 2015.
Fenollosa, Ernest, and Ezra Pound. "The Chinese Written Character as a Medium for Poetry: An Ars Poetica (1918–1936)." In *The Chinese Written Character as a Medium for Poetry: A Critical Edition*, edited by Haun Saussy, Jonathan Stalling, and Lucas Klein, 41–61. New York: Fordham University Press, 2008.
Feuer, Lewis S. "American Travelers to the Soviet Union 1917–32: The Formation of a Component of New Deal Ideology." *American Quarterly* 14, no. 2 (1962): 119–149.
Fiks, Arsen P. "Alexander L. Tchijevsky: 'Exquisite Poet-Philosopher,' 'Creator of New Sciences,' or 'Charlatan.'" *Caduceus* 10 (1994): 179–185.
Filene, Peter G. *Americans and the Soviet Experiment, 1917–1933*. Cambridge, MA: Harvard University Press, 1967.
Finlay, John L. *Social Credit: The English Origins*. Montreal: McGill–Queens University Press, 1972.
Fischer, Louis. "Russia's New Revolution." *Nation* 130, no. 3376 (March 19, 1930): 322–323.
Fisher, Irving. *Stabilised Money: A History of the Movement*. London: George Allen & Unwin, 1935.
Fitzgerald, F. Scott. *The Beautiful and Damned*. 1922. Reprint, London: Penguin Books, 2004.
Fitzhugh Talman, C. "Science Studies Effects of Sunspots on Earth: Their Influence on Ultra-violet Radiation." *New York Times*, July 17, 1927, xx4.
Fleming, James Rodger. *Historical Perspectives on Climate Change*. Oxford: Oxford University Press, 1998.

Fox, Richard Wightman. "Tragedy, Responsibility, and the American Intellectual, 1925–1950." In *Lewis Mumford: Public Intellectual*, edited by Thomas Parke Hughes and Agatha C. Hughes, 323–337. Oxford: Oxford University Press, 1990.

Frank, Waldo. "Our Guilt in Fascism." *New Republic* 102 (May 6, 1940): 603–608.

Franzese, Sergio. *The Ethics of Energy: William James's Moral Philosophy in Focus*. Frankfurt: Ontos, 2008.

Franzese, Sergio. "James versus Nietzsche: Energy and Asceticism in James." *Streams of William James* 5, no. 2 (2003): 10–12.

Frazer, Harold. "The Energy Certificate." *Technocracy Inc.* A, no. 10 (July 1937): 11–13.

Frederick, Christine. "The Psychology of Women under Technocracy." In *For and Against, Technocracy, a Symposium*, edited by J. G. Frederick, 109–131. New York: Business Bourse, 1933.

French, Daniel. *When They Hid the Fire: A History of Electricity and Invisible Energy in America*. Pittsburgh: University of Pittsburgh Press, 2017.

Freud, Sigmund. "Beyond the Pleasure Principle." In *Standard Edition of the Complete Psychological Works of Sigmund Freud*, vol. 18, *(1920–1922): Beyond the Pleasure Principle, Group Psychology, and Other Works*, edited by James Strachey et al., 7–64. London: Hogarth, 1955.

Freud, Sigmund. "Civilization and Its Discontents." In *The Standard Edition of the Complete Psychological Works of Sigmund Freud*, vol. 21, *(1927–1931)*, edited by James Strachey et al., 59–145. London: Hogarth, 1961.

Freud, Sigmund. "The Ego and the Id." In *The Standard Edition of the Complete Psychological Works of Sigmund Freud*, vol. 19, *(1923–25): The Ego and the Id, and Other Works*, edited by James Strachey et al., 12–66. London: Hogarth, 1962.

Freund, Daniel. *American Sunshine: Diseases of Darkness and the Quest for Natural Light*. Chicago: University of Chicago Press, 2012.

Friedman, Lawrence J., and Anke M. Schreiber. *The Lives of Erich Fromm: Love's Prophet*. New York: Columbia University Press, 2013.

Fromm, Erich. *The Fear of Freedom*. London: Routledge & Kegan Paul, 1942.

Fuller, R. Buckminster. *Ideas and Integrities: A Spontaneous Autobiographical Disclosure*. Englewood Cliffs, NJ: Prentice Hall, 1963.

Fuller, R. Buckminster. *Nine Chains to the Moon*. 1938. Reprint, Carbondale: Southern Illinois University Press, 1963.

Funk, Rainer. *Erich Fromm: His Life and Ideas. An Illustrated Biography*. New York: Continuum, 2000.

Galison, Peter, Gerald Holton, and Silvan S. Schweber. *Einstein for the 21st Century: His Legacy in Science, Art, and Modern Culture*. Princeton, NJ: Princeton University Press, 2008.

Garcia-Mata, Carlos, and Felix I. Shaffner. "Solar and Economic Relationships: A Preliminary Report." *Quarterly Journal of Economics* 49, no. 1 (1934): 1–51.

Gardiner, Rolf. *World without End: British Politics and the Younger Generation*. London: Cobden-Sanderson, 1932.

Gascoigne, Robert M. "Julian Huxley and Biological Progress." *Journal of the History of Biology* 24, no. 3 (1991): 433–455.

Gasman, Daniel. *The Scientific Origins of National Socialism: Social Darwinism in Ernst Haeckel and the German Monist League*. London: Macdonald, 1971.

Gauquelin, Michel. *The Cosmic Clocks: From Astrology to a Modern Science*. Chicago: H. Regnery, 1967.
Geddes, Patrick. *Cities in Evolution*. 1915. Reprint, London: Williams & Norgate, 1949.
Geddes, Patrick. *John Ruskin, Economist*. Edinburgh: William Brown, 1884.
Genter, Robert. *Late Modernism: Art, Culture, and Politics in Cold War America*. Philadelphia: University of Pennsylvania Press, 2010.
"Germany Has Been Prussianized, Says Bergson: Famous French Philosopher Declares Admiration of Brute Force Has Become a National Ideal." *New York Times*, January 10, 1915, 43.
Ghosh, Amitav. "Petrofiction: The Oil Encounter and the Novel." *New Republic*, March 2, 1992, 29–34.
Gilfillan, S. C. "The Coldward Course of Progress." *Political Science Quarterly* 35, no. 3 (1920): 393–410.
Glock, Waldo S. "The Language of Tree Rings." *Scientific Monthly* 38, no. 6 (1934): 501–510.
Gold, Barri J. *ThermoPoetics: Energy in Victorian Literature and Science*. Cambridge, MA: MIT Press, 2010.
Gooday, Graeme. *Domesticating Electricity: Technology, Uncertainty, and Gender, 1880–1914*. London: Pickering & Chatto, 2008.
Gooday, Graeme. "Sunspots, Weather, and the Unseen Universe: Balfour Stewart's Anti-materialist Representations of 'Energy' in British Periodicals." In *Science Serialized: Representations of the Sciences in Nineteenth-Century Periodicals*, edited by Geoffrey Cantor and Sally Shuttleworth, 111–147. Cambridge, MA: MIT Press, 2004.
Goodbody, Axel, and Brandon Smith. "Stories of Energy: Narrative in the Energy Humanities." *Resilience: A Journal of the Environmental Humanities* 6, no. 2 (2019): 1–25.
Goodman, Paul. "The Political Meaning of Some Recent Revisions of Freud." *politics* 2, no. 7 (July 1945): 197–203.
Goodman, Paul. "Revolution, Sociolatry and War." *politics* 2, no. 12 (December 1945): 376–380.
Gosling, Francis G. *Before Freud: Neurasthenia and the American Medical Community, 1870–1910*. Champaign: University of Illinois Press, 1987.
Goux, Jean-Joseph. *Symbolic Economies: After Marx and Freud*. Translated by Jennifer Curtiss Gage. Ithaca, NY: Cornell University Press, 1990.
Graetz, Michael J. *The End of Energy: The Unmaking of America's Environment, Security, and Independence*. Cambridge, MA: MIT Press, 2011.
"Gran'pa's Right! The Record Proves That Winters Are Getting Warmer." *Atlanta Constitution*, April 20, 1939, 1.
Gray, G. "Tornadoes on the Sun." *Popular Mechanics* 53, no. 4 (April 1930): 554–558.
Greer, Guy. "The Russian Bear in Business." *Outlook* 137, no. 2 (January 14, 1931): 55–59, 75–79.
Greif, Mark. *The Age of the Crisis of Man: Thought and Fiction in America, 1933–1973*. Princeton, NJ: Princeton University Press, 2015.
Griffith, Gareth. *Socialism and Superior Brains: The Political Thought of Bernard Shaw*. New York: Routledge, 1993.
Hagerty, James A. "Portland Cheers Speech." *New York Times*, September 22, 1932, 1.

Hale, Nathan G. "From Bergstrasse XIX to Central Park West: The Americanization of Psychoanalysis 1919–1940." *Journal of the History of the Behavioral Sciences* 14 (1974): 299–315.

Hale, Nathan G. *The Rise and Crisis of Psychoanalysis in the United States: Freud and the Americans, 1917–1985*. Oxford: Oxford University Press, 1995.

Hale, Piers J. "The Search for Purpose in a Post-Darwinian Universe: George Bernard Shaw, 'Creative Evolution,' and Shavian Eugenics: 'The Dark Side of the Force.'" *History and Philosophy of the Life Sciences* 28, no. 2 (2006): 191–213.

Halliday, Sam. *Science and Technology in the Age of Hawthorne, Melville, Twain, and James: Thinking and Writing Electricity*. New York: Palgrave Macmillan, 2007.

Hapke, Thomas. "Wilhelm Ostwald, Die Brücke (Bridge), and Connections to Other Bibliographic Activities at the Beginning of the Twentieth Century." In *Proceedings of the 1998 Conference on the History and Heritage of Science Information Systems*, edited by Mary Ellen Bowden, Trudi Bellardo Hahn, and Robert Virgil Williams, 139–147. Medford: Information Today, 1999.

Harrison-Moore, Abigail, and Ruth Sandwell. *In a New Light: Histories of Women and Energy*. Montreal: McGill–Queens University Press, 2021.

Hayek, Friedrich A. von. *The Counter-Revolution of Science: Studies on the Abuse of Reason*. Glencoe, IL: Free Press, 1952.

Hayek, Friedrich A. von. *The Road to Serfdom*. Chicago: University of Chicago Press, 1944.

Hayles, Katherine N. *Chaos Bound: Orderly Disorder in Contemporary Literature and Science*. Ithaca, NY: Cornell University Press, 1990.

Hayles, Katherine N. *How We Became Posthuman: Virtual Bodies in Cybernetics, Literature, and Informatics*. Chicago: University of Chicago Press, 1999.

Hays, Samuel P. *Conservation and the Gospel of Efficiency: The Progressive Conservation Movement, 1890–1920*. Pittsburgh: University of Pittsburgh Press, 1999.

Hazlitt, Henry. "How High Is Up? The Fallacy of Absolute Standards." *Forum* 87, no. 2 (February 1932): 115–119.

Hazlitt, Henry. "Scrambled Ergs: An Examination of Technocracy." *Nation* 136, no. 3526 (February 1, 1933): 112–115.

"Head of TVA Denies Utilities' Charges: Lilienthal Insists before House Group That Project Is Not a 'Rubber Yardstick.'" *New York Times*, April 5, 1935, 39.

"Heat and Population." *New York Times*, February 6, 1939, 8.

"Heat Belt Seen Shifting Northward by Professor." *Evening Independent* (St. Petersburg, Florida), July 28, 1936, 1.

Heidelberger, Michael. *Nature from Within: Gustav Theodor Fechner and His Psychophysical Worldview*. Pittsburgh: University of Pittsburgh Press, 2004.

Henriksen, Margot A. *Dr. Strangelove's America Society and Culture in the Atomic Age*. Berkeley: University of California Press, 1997.

"Herr Professor Ostwald Wants to Organize a Brain Trust." *Milwaukee Journal*, March 9, 1913, 8.

Hitchcock, Peter. "Oil in an American Imaginary." *New Formations* 69, no. 3 (2010): 81–97.

Hocking, William Ernest. "The Atom as Moral Dictator." *Saturday Review of Literature* 29, no. 5 (February 2, 1946): 7–9.

Hoganson, Kristin L. *Fighting for American Manhood: How Gender Politics Provoked the Spanish-American and Philippine-American Wars.* New Haven, CT: Yale University Press, 1998.

Hoh, E. H. J. "Christianity and Technocracy." *Mercury*, January 31, 1933, 2.

Hoke, Travis. "Utopia by Thermometer." *North American Review* 234, no. 2 (August 1932): 110–116.

Holt, Niles R. "Wilhelm Ostwald's 'The Bridge.'" *British Journal for the History of Science* 10, no. 2 (July 1, 1977): 146–150.

Holton, Gerald. "Ernst Mach and the Fortunes of Positivism in America." *Isis* 83, no. 1 (March 1, 1992): 27–60.

Hook, Sidney. "The New Failure of Nerve." *Partisan Review* 10, no. 1 (January–February 1943): 2–23.

Horkheimer, Max, and Theodor W. Adorno. *Dialectic of Enlightenment.* New York: Herder & Herder, 1972.

Horowitz, Daniel. *The Anxieties of Affluence: Critiques of American Consumer Culture, 1939–1979.* Amherst: University of Massachusetts Press, 2004.

Huber, Matthew. T. *Lifeblood: Oil, Freedom, and the Forces of Capital.* Minneapolis: University of Minnesota Press, 2013.

Hughes, Thomas P. *Networks of Power: Electrification in Western Society, 1880–1930.* Baltimore: Johns Hopkins University Press, 1983.

Huntington, Ellsworth. "Babbitt and the Weather." *Outlook* 153, no. 5 (October 2, 1929): 186.

Huntington, Ellsworth. "Changes of Climate and History." *American Historical Review* 18, no. 2 (1913): 213–232.

Huntington, Ellsworth. *Civilization and Climate.* New Haven, CT: Yale University Press, 1915.

Huntington, Ellsworth. "Climatic Cycles." In *Matamek Conference on Biological Cycles*, together with a report by Ellsworth Huntington, 142–144. Canadian Labrador: Matamek Factory, 1932.

Huntington, Ellsworth. "The Ebb and Flow of Human Populations." In *Matamek Conference on Biological Cycles*, together with a report by Ellsworth Huntington, 60–69. Canadian Labrador: Matamek Factory, 1932.

Huntington, Ellsworth. *Mainsprings of Civilization.* New York: Wiley & Sons, 1945.

Huntington, Ellsworth. "The Matamek Conference on Biological Cycles, 1931." *Science* 74, no. 1914 (1931): 229–235.

Huntington, Ellsworth. *The Pulse of Asia: A Journey in Central Asia Illustrating the Geographic Basis of History.* New York: Houghton Mifflin, 1907.

Huntington, Ellsworth. "Russia's Worst Enemy." *New Outlook* 161, no. 7 (April 1933): 36–39.

Huntington, Ellsworth. "The Social Pyramid: *Social Mobility*, by Pitirim A. Sorokin." *Saturday Review* 4, no. 19 (December 3, 1927): 370–371.

Huntington, Ellsworth. *World-Power and Evolution.* New Haven, CT: Yale University Press, 1919.

Huntington, Ellsworth, and Henry Helm Clayton. *Earth and Sun: An Hypothesis of Weather and Sunspots.* New Haven, CT: Yale University Press, 1923.

Huntington, Ellsworth, and Stephen Sargent Visher. *Climatic Changes, Their Nature and Causes*. New Haven, CT: Yale University Press, 1922.

Huxley, Aldous. *The Doors of Perception*. New York: Harper & Brothers, 1954.

Huxley, Aldous. *Ends and Means: An Inquiry into the Nature of Ideals and into Methods Employed for Their Realization*. New York: Harper & Brothers, 1937.

Huxley, Aldous. *Eyeless in Gaza*. 1936. Reprint, New York: American Reprint Company, 1975.

Huxley, Aldous. Introduction to *Hopousia; or, The Sexual and Economic Foundations of a New Society*, by J. D. Unwin, 13–29. London: George Allen & Unwin, 1940.

Huxley, Julian. "The Crisis in Man's Destiny." *Playboy* 14 (January 1967): 94, 212–217.

Huxley, Julian. *Evolutionary Ethics*. Oxford: Oxford University Press, 1943.

Huxley, Julian. *If I Were Dictator*. New York: Harper & Brothers, 1934.

Huxley, Julian. Introduction to *The Phenomenon of Man*, by Pierre Teilhard de Chardin, 11–28. New York: Harper, 1959.

Huxley, Julian. "On Living in a Revolution." *Harper's Magazine* 185, no. 1108 (1942): 337–347.

Huxley, Julian. *On Living in a Revolution*. New York: Harper, 1944.

Huxley, Julian. *Religion without Revelation*. 1927. Reprint, New York: Harper & Brothers, 1957.

Huxley, Julian. "Science, War and Reconstruction." *Science* 91, no. 2355 (1940): 151–158.

Huxley, Julian. *A Scientist among the Soviets*. New York: Harper & Brothers, 1932.

Huxley, Julian. *TVA, Adventure in Planning*. Surrey: Architectural Press, 1943.

Huxley, Julian. "UNESCO: Its Purpose and Its Philosophy." *Free World* 7, no. 5 (December 1946): 27–30.

Huxley, Julian. *UNESCO: Its Purpose and Its Philosophy*. Washington, DC: Public Affairs Press, 1947.

Huxley, Thomas Henry. *Evolution and Ethics and Other Essays*. London: Macmillan, 1895.

"Industrial Growth of Nation Is Traced: 'Energy Survey' Undertaken by Columbia University to Cover the Last 100 Years." *New York Times*, August 6, 1932, 13.

"Is Our Civilization Doomed? Explaining the Death of Other Cultures, Mr. Toynbee Holds Out Hope for Our Own." *New York Times*, April 13, 1947, 203.

"Jackson Pollock, Is He the Greatest Living Painter in the United States?" *Life* 27, no. 6 (August 8, 1949): 42–45.

Jaffe, Aaron. "Inventing the Radio Cosmopolitan: Vernacular Modernism at a Standstill." In *Broadcasting Modernism*, edited by Debra Rae Cohen, Michael Coyle, and Jane Lewty, 11–32. Tallahassee: University of Florida Press, 2009.

James, Henry. *The Letters of William James*. Vol. 2. Boston: Atlantic Monthly Press, 1920.

James, William. "The Energies of Men" (1906). In *Writings, 1902–1910*, edited by B. Kuklick, 1223–1241. New York: Literary Classics of the United States, 1987.

James, William. *The Energies of Men*. New York: Moffat, Yard, 1914.

James, William. "Herbert Spencer." *Atlantic Monthly* 94 (July 1904): 99–108.

James, William. "The Moral Equivalent of War" (1910). In *William James, Writings, 1902–1910*, edited by B. Kuklick, 1281–1293. New York: Literary Classics of the United States, 1987.

James, William. "Pragmatism's Conception of Truth" (1907). In *Pragmatism, and Four Essays from "The Meaning of Truth,"* 131–157. Cleveland: Meridian Books, 1942.

James, William. *The Principles of Psychology*. 1890. Reprint, Cambridge, MA: Harvard University Press, 1981.

James, William. *The Varieties of Religious Experience*. 1902. Reprint, Cambridge, MA: Harvard University Press, 1985.

Jay, Martin. *The Dialectical Imagination: A History of the Frankfurt School and the Institute of Social Research, 1923–1950*. Berkeley: University of California Press, 1973.

Jefferies, Matthew, and Mike Tyldesley. *Rolf Gardiner: Folk, Nature and Culture in Interwar Britain*. Farnham: Ashgate, 2011.

Jeffries, John W. "The 'Quest for National Purpose' of 1960." *American Quarterly* 30, no. 4 (1978): 191.

Jevons, H. Stanley. "The Causes of Fluctuations of Industrial Activity and the Price-Level." *Journal of the Royal Statistical Society* 96, no. 4 (1933): 545–605.

Jewett, Andrew. *Science, Democracy, and the American University: From the Civil War to the Cold War*. Cambridge: Cambridge University Press, 2012.

Johnson, Benjamin Heber. *Escaping the Dark, Gray City: Fear and Hope in Progressive-Era Conservation*. New Haven, CT: Yale University Press, 2017.

Johnson, Bob. *Carbon Nation: Fossil Fuels in the Making of American Culture*. Lawrence: University Press of Kansas, 2014.

Johnson, Bob. *Mineral Rites: An Archaeology of the Fossil Economy*. Baltimore: Johns Hopkins University Press, 2019.

Johnson, Gerald W. "Bryan, Thou Shouldst Be Living." *Harper's Monthly Magazine* 163, no. 976 (September 1931): 385–391.

Jones, Chris F. "Petromyopia: Oil and the Energy Humanities." *Humanities* 5, no. 2 (2016): 1–10.

Jones, Chris F. *Routes of Power: Energy and Modern America*. Cambridge, MA: Harvard University Press, 2014.

Jordan, John M. *Machine-Age Ideology: Social Engineering and American Liberalism, 1911–1939*. Chapel Hill: University of North Carolina Press, 1994.

Jordan, Virgil. "Technocracy—Tempest on a Slide Rule." *Scribner's Magazine* 93, no. 2 (1933): 65–69.

Jung, Carl Gustav. "On Psychic Energy." In *The Collected Works of C. G. Jung: The Structure and Dynamics of the Psyche*, edited by Gerhard Adler and R. F. C Hull, 3–67. London: Routledge & Kegan Paul, 1960.

Kaempffert, Waldemar. "The Sun: Mysterious Source of All Life and Energy." *New York Times*, August 16, 1931, 106.

Kander, Astrid, Paolo Malanima, and Paul Warde. *Power to the People: Energy in Europe over the Last Five Centuries*. Princeton, NJ: Princeton University Press, 2018.

Karmel, Pepe, and Kirk Varnedoe. *Jackson Pollock: Interviews, Articles, and Reviews*. New York: Museum of Modern Art, 1999.

Katznelson, Ira. *Fear Itself: The New Deal and the Origins of Our Time*. New York: Liveright, 2013.

Keller, Morton. *Regulating a New Economy: Public Policy and Social Change in America, 1900–1933*. Cambridge, MA: Harvard University Press, 1994.

Kennedy, Dane. "The Perils of the Midday Sun: Climatic Anxieties in the Colonial Tropics." In *Imperialism and the Natural World*, edited by John M. MacKenzie, 118–140. Manchester: Manchester University Press, 1990.

Kennedy, David M. *Freedom from Fear: The American People in Depression and War, 1929–1945*. Oxford: Oxford University Press, 1999.

Kern, Stephen. *The Culture of Time and Space, 1880–1918*. Cambridge, MA: Harvard University Press, 2003.

Keun, Odette. *A Foreigner Looks at the TVA*. New York: Longmans, Green, 1937.

Killen, Andreas. *Berlin Electropolis: Shock, Nerves, and German Modernity*. Berkeley: University of California Press, 2006.

Kincer, J. B. "Is Our Climate Changing to Milder?" *Scientific Monthly* 39, no. 1 (July 1934): 59–62.

Kuhn, Thomas S. "Energy Conservation as an Example of Simultaneous Discovery." In *Critical Problems in the History of Science*, edited by Marshall Clagett, 321–356. Madison: University of Wisconsin Press, 1959.

Labanca, Nicola. "Ontological Fallacies Linked to Energy, Information and Related Technologies." In *Complex Systems and Social Practices in Energy Transitions: Framing Energy Sustainability in the Time of Renewables*, ed. Nicola Labanca, 171–206. Cham, Switzerland: Springer International, 2017.

LaFeber, Walter. *The New Empire: An Interpretation of American Expansion, 1860–1898*. Ithaca, NY: Cornell University Press, 1998.

LaFollette, Marcel C. *Science on the Air: Popularizers and Personalities on Radio and Early Television*. Chicago: University of Chicago Press, 2008.

Land, Mary G. "Three Max Gottliebs: Lewis's, Dreiser's, and Walker Percy's View of the Mechanist-Vitalist Controversy." *Studies in the Novel* 15, no. 4 (1983): 314–331.

Lane, Richard. "The Nature of Growth: The Postwar History of the Economy, Energy and the Environment." PhD diss., University of Sussex, 2015.

Lasch, Christopher. *The True and Only Heaven: Progress and Its Critics*. New York: Norton, 1991.

Lawrence, D. H. *Studies in Classic American Literature*. 1923. Reprint, Cambridge: Cambridge University Press, 2003.

Lawrence, D. H. "The Sun" (1925/1928). In *The Cambridge Edition of the Works of D. H. Lawrence: The Woman Who Rode Away and Other Stories*, edited by Dieter Mehl and Christa Jansohn, 19–38. Cambridge: Cambridge University Press, 1995.

Leach, Henry Goddard. "Education for Patriotism." *Forum* 93, no. 5 (May 1935): 257–258.

Lears, T. J. Jackson. *No Place of Grace: Antimodernism and the Transformation of American Culture, 1880–1920*. Chicago: University of Chicago Press, 1981.

Lee, Martin A., and Bruce Shlain. *Acid Dreams: The Complete Social History of LSD: The CIA, the Sixties, and Beyond*. New York: Grove Weidenfeld, 1992.

Leja, Michael. *Reframing Abstract Expressionism: Subjectivity and Painting in the 1940s*. New Haven, CT: Yale University Press, 1993.

LeMenager, Stephanie. *Living Oil: Petroleum Culture in the American Century*. New York: Oxford University Press, 2016.

Leopold, Aldo. "Wilderness as a Form of Land Use." *Journal of Land and Public Utility Economics* 1, no. 4 (October 1, 1925): 398–404.

"Letters to the Editor." *Saturday Review* 28, no. 36 (September 8, 1945): 19–21.

Leuchtenburg, William E. *Franklin D. Roosevelt and the New Deal: 1932–1940*. New York: Harper & Row, 1963.
Levine, Daniel. "Randolph Bourne, John Dewey and the Legacy of Liberalism." *Antioch Review* 29, no. 2 (July 1, 1969): 234–244.
Lewis, Michael L. *American Wilderness: A New History*. Oxford: Oxford University Press, 2007.
Lieberman, Jennifer. *Power Lines: Electricity in American Life and Letters, 1882–1952*. Cambridge: Cambridge University Press, 2017.
Lilienthal, David E. *TVA—Democracy on the March*. New York: Harper & Brothers, 1944.
Lindeman, Eduard. "Morality for the Atomic Age." *Forum* 106, no. 3 (November 1, 1945): 231–233.
Lippmann, Walter. *A Preface to Politics*. 1913. Reprint, Ann Arbor: University of Michigan Press, 1962.
Lippmann, Walter. "Technocracy: The Prophecy of Doom." *Boston Globe*, February 2, 1933, 14.
Lippmann, Walter. "The World Conflict in Relation to American Democracy." *Annals of the American Academy of Political and Social Science* 72 (1917): 1–10.
Livingston, James. "War and the Intellectuals: Bourne, Dewey, and the Fate of Pragmatism." *Journal of the Gilded Age and Progressive Era* 2, no. 4 (2003): 431–450.
Livingstone, D. N. "Darwinian Hippocratics, Eugenic Enticements, and the Biometeorological Body." In *Weather, Climate and the Geographical Imagination*, edited by M. Mahony and S. Randalls. Pittsburgh: University of Pittsburgh Press, 2020, 191–214.
Loeb, Harold. *Life in a Technocracy, What It Might Be Like*. Edited by Howard P. Segal. 1933. Reprint, Syracuse: Syracuse University Press, 1996.
Loeb, Jacques. "Freedom of Will and War." *New Review* 2, no. 11 (1914): 631–636.
Loeb, Jacques. "The Mechanistic Conception of Life" (1912). In *The Mechanistic Conception of Life*, 1–31. Cambridge, MA: Belknap Press of Harvard University Press, 1964.
Loeb, Jacques. "Mechanistic Science and Metaphysical Romance." *Yale Review* 4 (July 1915): 768–785.
Loeb, Jacques. "The Significance of Tropisms for Psychology" (1909). In *The Mechanistic Conception of Life*, 33–62. Cambridge, MA: Belknap Press of Harvard University Press, 1964.
López, Rick Anthony. *Crafting Mexico: Intellectuals, Artisans, and the State after the Revolution*. Durham, NC: Duke University Press, 2010.
Luccarelli, Mark. *Lewis Mumford and the Ecological Region: The Politics of Planning*. New York: Guilford, 1996.
Lutz, Tom. *American Nervousness, 1903: An Anecdotal History*. Ithaca, NY: Cornell University Press, 1991.
M., C. "A Moral Equivalent for the Red Hysteria." *New Republic* 21, no. 272 (February 18, 1920): 345–346.
Macdonald, Graeme. "Research Note: The Resources of Fiction." *Reviews in Cultural Theory* 4, no. 2 (2013): 1–24.
MacDuffie, Allen. *Victorian Literature: Energy, and the Ecological Imagination*. Cambridge: Cambridge University Press, 2014.
Mach, Ernst. *History and Root of the Principle of the Conservation of Energy*. Translated by Philip E. B. Jourdain. Chicago: Open Court, 1911.

Mach, Ernst. *Principles of the Theory of Heat: Historically and Critically Elucidated.* Edited by Brian McGuinness. 1896. Reprint, Dordrecht: D. Reidel, 1986.

MacKaye, Benton. "Appalachian Power: Servant or Master." *Survey Graphic* 51, no. 11 (March 1, 1924): 618–619.

MacKaye, Benton. "An Appalachian Trail: A Project in Regional Planning." *Journal of the American Institute of Architects* 9 (October 1921): 325–330.

MacKaye, Benton. "End or Peak of Civilisation?" *Survey Graphic* 68, no. 13 (October 1, 1932): 441–444.

MacKaye, Benton. *From Geography to Geotechnics.* Edited by Paul T. Bryant. Champaign: University of Illinois Press, 1968.

MacKaye, Benton. *The New Exploration: A Philosophy of Regional Planning.* 1928. Reprint, Champaign: University of Illinois Press, 1962.

MacKaye, Benton. "The New Exploration: Charting the Industrial Wilderness." *Survey Graphic* 54, no. 3 (May 1, 1925): 153–157.

MacKaye, Benton. "The Tennessee River Project: First Step in a National Plan." *New York Times*, April 16, 1933, xx3.

MacKaye, Benton. "Tennessee—Seed of a National Plan." *Survey Graphic* 22, no. 5 (May 1933): 251–254.

MacKaye, Benton. "War and Wilderness." *Living Wilderness* 6 (July 1941): 7–8.

MacLeish, Archibald. "If We Want Peace, This Is the First Job: We Must Begin by Trying to Create a Real Sense of a World Community." *New York Times*, November 17, 1946, 11, 60–61.

MacLeish, Archibald. *The Irresponsibles: A Declaration.* New York: Duell, Sloan & Pearce, 1940.

MacLeish, Archibald. "Machines and the Future." *Nation* 136, no. 3527 (1933): 140–142.

Maher, Neil M. *Nature's New Deal: The Civilian Conservation Corps and the Roots of the American Environmental Movement.* Oxford: Oxford University Press, 2008.

Malinowski, Bronislaw. *Sex and Repression in Savage Society.* 1927. Reprint, London: Routledge, 2003.

Malm, Andreas. *Fossil Capital: The Rise of the Steam Engine and the Roots of Global Warming.* London: Verso, 2016.

Malm, Andreas, and Alf Hornborg. "The Geology of Mankind? A Critique of the Anthropocene Narrative." *Anthropocene Review* 1, no. 1 (2014): 62–69.

Malone, George W. "Power—Who Gets It?" *Michigan Technic* 49, no. 6 (March 1936): 5–7.

"Man Is Slave of Sun-Spots Scientist Says." *Courier-Post*, December 31, 1926, 1.

Mansfield, Nick. *The God Who Deconstructs Himself: Sovereignty and Subjectivity between Freud, Bataille, and Derrida.* New York: Fordham University Press, 2010.

"Man Slave of Sun, Avers Scientist." *Brooklyn Daily Times*, December 31, 1929, 1.

Marshall, Robert. *Arctic Village.* London: Jarrolds, 1934.

Martin, Edward. "Editor's Easy Chair: Sun Spots and Justice." *Harper's Monthly Magazine* 155, no. 930 (November 1927): 785–788.

Martin, Edward. "Editor's Easy Chair: Sun Spots and Politics." *Harper's Monthly Magazine* 155, no. 923 (March 1927): 529–532.

Martin, Geoffrey J. *American Geography and Geographers: Toward Geographical Science.* Oxford: Oxford University Press, 2015.

Martin, Geoffrey J. *Ellsworth Huntington: His Life and Thought*. Hamden, CT: Archon Books, 1973.

Martin, Ronald E. *American Literature and the Universe of Force*. Durham, NC: Duke University Press, 1981.

Martinez-Alier, Joan. *Ecological Economics: Energy, Environment, and Society*. Oxford: Basil Blackwell, 1987.

Marx, Leo. "Lewis Mumford: Prophet of Organicism." In *Lewis Mumford: Public Intellectual*, edited by Thomas P. Hughes and Agatha C. Hughes, 164–180. Oxford: Oxford University Press, 1990.

Marx, Leo. *The Machine in the Garden: Technology and the Pastoral Ideal in America*. Oxford: Oxford University Press, 1964.

"Mass Excitement Laid to Sunspots; Dr V. P. De Smitt Says Russian Scientist Has Traced World Unrest to Solar Phenomena." *New York Times*, March 8, 1927, 52.

Massey, Jonathan. "Necessary Beauty: Fuller's Sumptuary Aesthetics." In *New Views on R. Buckminster Fuller*, edited by Hsiao-Yun Chu and Roberto G. Trujillo, 99–124. Stanford, CA: Stanford University Press, 2009.

Matamek Conference on Biological Cycles. Full Proceedings. Together with a Report by Ellsworth Huntington. Canadian Labrador: Matamek Factory, 1932.

Matless, David. "Bodies Made of Grass Made of Earth Made of Bodies: Organicism, Diet and National Health in Mid-twentieth-century England." *Journal of Historical Geography* 27, no. 3 (2001): 355–376.

Matless, David. *Landscape and Englishness*. London: Reaktion Books, 1998.

Matless, David. "Regional Surveys and Local Knowledges: The Geographical Imagination in Britain, 1918–39." *Transactions of the Institute of British Geographers* 17, no. 4 (1992): 464–480.

Mattson, Kevin. *Intellectuals in Action: The Origins of the New Left and Radical Liberalism, 1945–1970*. University Park: Pennsylvania State University Press, 2002.

McClay, Wilfred M. *The Masterless: Self and Society in Modern America*. Chapel Hill: University of North Carolina Press, 1994.

McCraw, Thomas K. *Morgan vs. Lilienthal: The Feud within the TVA*. Chicago: Loyola University Press, 1970.

McNeill, John Robert, and Peter Engelke. *The Great Acceleration: An Environmental History of the Anthropocene since 1945*. Cambridge, MA: Harvard University Press, 2014.

"The Mechanistic Conception of Life." *Athenaeum*, October 19, 1912.

"The Mechanistic Conception of Life." *Boston Transcript*, November 8, 1912.

"The Mechanistic Conception of Life." *Chicago Evening Post*, February 28, 1913.

"The Mechanistic Conception of Life." *Daily People*, November 19, 1912.

"The Mechanistic Conception of Life." *Hebrew Standard*, July 11, 1913.

"The Mechanistic Conception of Life." *Inquirer*, December 28, 1912.

"The Mechanistic Conception of Life." *Knowledge*, November 1912.

"The Mechanistic Conception of Life." *London Times*, September 19, 1912.

"The Mechanistic Conception of Life." *Nation*, December 5, 1912.

"The Mechanistic Conception of Life." *San Francisco Chronicle*, September 15, 1912.

"The Mechanistic Conception of Life." *Scotsman*, October 28, 1912.

Meissner, B. F. "The Electric Dog and How He Obeys His Flash-Lamp Master." *Popular Science* 88, no. 3 (March 1916): 426–429.

Meller, Helen Elizabeth. *Patrick Geddes: Social Evolutionist and City Planner*. London: Routledge, 1990.

Mendelsohn, Everett I. "Prophet of Our Discontent: Lewis Mumford Confronts the Bomb." In *Lewis Mumford: Public Intellectual*, edited by Thomas P. Hughes and Agatha C. Hughes, 343–360. Oxford: Oxford University Press, 1990.

Merchant, Carolyn. *Reinventing Eden: The Fate of Nature in Western Culture*. London: Routledge, 2013.

Merricks, Linda. *The World Made New: Frederick Soddy, Science, Politics, and Environment*. Oxford: Oxford University Press, 1996.

Meyer, Frank. "Long-Term Hope for Humanity, UNESCO: Its Purpose and Its Philosophy." *New York Times*, September 7, 1947, BR32.

Miller, Donald L. *Lewis Mumford, A Life*. New York: Weidenfeld & Nicolson, 1989.

Mills, Clarence A. *Climate Makes the Man*. New York: Harper & Brothers, 1942.

Mills, Clarence A. *Living with the Weather*. Cincinnati: Caxton, 1934.

Mills, Clarence A. "Oncoming Reversal of the Human Growth Tide." *Science* 92, no. 2392 (1940): 401–402.

Mills, Clarence A. "They Mature *Later* in the Tropics." *Harper's Magazine* 185 (February 1, 1942): 294–298.

Mills, C. Wright, and Patricia J. Salter. "The Barricade and the Bedroom." *politics* 2, no. 10 (October 1945): 313–315.

Minteer, Ben A. *The Landscape of Reform Civic Pragmatism and Environmental Thought in America*. Cambridge, MA: MIT Press, 2006.

Mirowski, Philip. *The Effortless Economy of Science?* Durham, NC: Duke University Press, 2004.

Mirowski, Philip. "Energy and Energetics in Economic Theory: A Review Essay." *Journal of Economic Issues* 22, no. 3 (1988): 811–830.

Mirowski, Philip. *More Heat Than Light: Economics as Social Physics, Physics as Nature's Economics*. Cambridge: Cambridge University Press, 1989.

"Misdirected Energies." *New York Times*, April 2, 1928, 16.

Mitchell, Timothy. *Carbon Democracy: Political Power in the Age of Oil*. London: Verso, 2013.

Molesworth, Charles. "Inner and Outer: The Axiology of Lewis Mumford." In *Lewis Mumford: Public Intellectual*, edited by Thomas P. Hughes and Agatha C. Hughes, 241–255. Oxford: Oxford University Press, 1990.

"Money Systems on Technocracy Weak Aver Economists." *Ellensburg Daily Record*, December 31, 1932, 1–2.

Montague, W. Pepperrell. "Are Mental Processes in Space?" *Monist* 18, no. 1 (January 1, 1908): 21–29.

Monteiro, George. "Henry Adams and William James: An Intellectual Relationship." *American Literary Realism* 39, no. 2 (2007): 150–167.

"Morality More Than an Instinct: Dr. Neumann Attacks Materialism." *Brooklyn Citizen*, December 9, 1912.

Morris, Kenneth E. *Jimmy Carter, American Moralist*. Athens: University of Georgia Press, 1996.

Morrison, Paul. *The Poetics of Fascism: Ezra Pound, T. S. Eliot, Paul de Man.* Oxford: Oxford University Press, 1996.
"The Most Baffling Fact of Human Knowledge: We May Have to Give up Matter and Form Conceptions or Be Unscientific." *Current Opinion* 65, no. 1 (July 1918): 35.
Mumford, Lewis. *The Condition of Man.* New York: Harcourt, Brace, 1944.
Mumford, Lewis. *The Conduct of Life.* New York: Harcourt, Brace, 1951.
Mumford, Lewis. "The Corruption of Liberalism." *New Republic* 102, no. 18 (April 29, 1940): 568–573.
Mumford, Lewis. "The Fourth Migration." *Survey Graphic* 54, no. 4 (May 1925): 130–133.
Mumford, Lewis. "Gentlemen: You Are Mad!" *Saturday Review of Literature* 29, no. 9 (March 2, 1946): 5–6.
Mumford, Lewis. *The Golden Day: A Study in American Experience and Culture.* New York: Boni & Liveright, 1926.
Mumford, Lewis. "Power and Culture." In *Third World Power Conference Transactions*, 1:167–172. Washington, DC: US Government Printing Office, 1936.
Mumford, Lewis. "Spengler: Dithyramb to Doom" (1939–1944). In *Interpretations and Forecasts, 1922–1972*, 218–224. New York: Harcourt Brace Jovanovich, 1973.
Mumford, Lewis. *Technics and Civilization.* New York: Harcourt, Brace, 1934.
Mumford, Lewis, and Patrick Geddes. *Lewis Mumford and Patrick Geddes: The Correspondence.* Edited by Frank G. Novak. London: Routledge, 1995.
"Muscle Shoals as a Yardstick for Electric Rates." *Literary Digest* 116, no. 14 (September 30, 1933): 9.
Myers, Greg. "Nineteenth-Century Popularizations of Thermodynamics and the Rhetoric of Social Prophecy." *Victorian Studies* 29, no. 1 (1985): 35–66.
Nadel, Alan. *Containment Culture: American Narrative, Postmodernism, and the Atomic Age.* Durham, NC: Duke University Press, 1995.
Nash, Roderick. *Wilderness and the American Mind.* New Haven, CT: Yale University Press, 1967.
Nash, Stephen E. "Time for Collaboration: A. E. Douglass, Archaeologists, and the Establishment of Tree-Ring Dating in the American Southwest." *Journal of the Southwest* 40, no. 3 (1998): 261–305.
"Nation's Newspaper Editorialists Split over Carter's Energy Talk." *New York Times*, July 17, 1979, A13.
Needham, Andrew. *Power Lines: Phoenix and the Making of the Modern Southwest.* Princeton, NJ: Princeton University Press, 2015.
Neuber, Matthias. "Uneasy Allies: Ostwald, Helm, Mach and Their Philosophies of Science." In *Wilhelm Ostwald at the Crossroads between Chemistry, Philosophy and Media Culture*, edited by Britta Görs, Nikos Psarros, and Paul Ziche, 47–58. Leipzig: Leipziger Universitätsverlag, 2005.
Neuse, Steven M. *David E. Lilienthal: The Journey of an American Liberal.* Knoxville: University of Tennessee Press, 1996.
"The New Realism—a Philosophy of Disillusionment." *Current Opinion*, January 1913, 52.
Nicholls, Peter. "Destinations, *Broom* (1921–4) and *Secession* (1922–4)." In *The Oxford Critical and Cultural History of Modernist Magazines*, vol. 2, *North America 1894–*

1960, edited by Peter Brooker and Andrew Thacker, 636–654. Oxford: Oxford University Press, 2009.

Nye, David E. *America's Assembly Line*. Cambridge, MA: MIT Press, 2013.

Nye, David E. *Consuming Power: A Social History of American Energies*. Cambridge, MA: MIT Press, 1998.

Nye, David E. *Electrifying America: Social Meanings of a New Technology, 1880–1940*. Cambridge, MA: MIT Press, 1990.

Nye, David E. "Energy in the Thought and Design of R. Buckminster Fuller." In *New Views on R. Buckminster Fuller*, edited by Hsiao-Yun Chu and Roberto G. Trujillo, 86–99. Stanford, CA: Stanford University Press, 2009.

Nye, David E. "Energy Narratives." *American Studies in Scandinavia* 25 (1993): 73–91.

Oliver, J. A. Westwood. *Sunspottery: Or, What Do We Owe to the Sun: A Popular Examination of the Cycle Theory of the Weather, Famines, Pestilences, Commercial Panics, etc.* London: Simpkin, Marshall, 1883.

Olson, Charles. "The Gate and the Center" (1951). In *Collected Prose: Charles Olson*, edited by Donald Allen and Benjamin Friedlander, 168–173. Berkeley: University of California Press, 1997.

Olson, Charles. *The Maximus Poems*. Edited by G. F. Butterick. Berkeley: University of California Press, 1983.

Olson, Charles. "Projective Verse" (1950). In *Collected Prose: Charles Olson*, edited by Donald Allen and Benjamin Friedlander, 239–249. Berkeley: University of California Press, 1997.

O'Neill, John. "Ecological Economics and the Politics of Knowledge: The Debate between Hayek and Neurath." *Cambridge Journal of Economics* 28, no. 3 (May 1, 2004): 431–447.

"One of Technocracy's Electricity Dollars Is Taken in Collection." *Ottawa Citizen*, January 9, 1933, 1.

Osborn, Henry Fairfield. *The Origin and Evolution of Life: On the Theory of Action, Reaction and Interaction of Energy*. New York: Charles Scribner's Sons, 1917.

Ostwald, Wilhelm. *Individuality and Immortality*. Boston: Houghton Mifflin, 1906.

Ostwald, Wilhelm. *Monism as the Goal of Civilization*. Hamburg: International Committee of Monism, 1913.

Ostwald, Wilhelm. "The Philosophical Meaning of Energy." *International Quarterly* 7 (March 1903): 300–315.

Ostwald, Wilhelm. "A Theory of Happiness." *International Quarterly* 11 (April 1, 1905): 316–326.

Overy, Paul. *Light, Air and Openness: Modern Architecture between the Wars*. London: Thames & Hudson, 2007.

Overy, Richard. *The Morbid Age: Britain between the Wars*. London: Allen Lane, 2009.

Painter, David S. "Oil in American History: A Special Issue." *Journal of American History* 99 (2012): 24–39.

Parker, Andrew. "Ezra Pound and the 'Economy' of Anti-Semitism." *Boundary 2* 11, no. 1/2 (October 1, 1982): 103–128.

Parrish, Wayne W. "Technocracy's Challenge." *New Outlook* 161, no. 4 (January 1933): 13–16.

Parrish, Wayne W. "Technocracy's Question." *New Outlook* 161, no. 3 (December 1932): 14.

Partington, John S. *Building Cosmopolis: The Political Thought of H. G. Wells.* Aldershot: Ashgate, 2003.
Paterson, Isabel. *The God of the Machine.* New York: G. P. Putnam's Sons, 1943.
Paterson, Isabel. "Turns with a Bookworm." *New York Herald Tribune "Books,"* September 1, 1933, 11.
Pauly, Philip J. *Controlling Life: Jacques Loeb and the Engineering Ideal in Biology.* New York: Oxford University Press, 1987.
Pearl, Raymond. "The Mechanistic Conception of Life." *Dial* 54, no. 638 (January 16, 1913): 51–52.
Pemberton, Jo-Anne. *Global Metaphors: Modernity and the Quest for One World.* London: Pluto, 2001.
Pemberton, Jo-Anne. "New Worlds for Old: The League of Nations in the Age of Electricity." *Review of International Studies* 28, no. 2 (2002): 311–336.
Peña, Carolyn de la. *The Body Electric: How Strange Machines Built the Modern American.* New York: New York University Press, 2005.
Phillips, H. I. "Technocracy Made Simple: Or, the Technocrat at the Tea Table." *Collier's Weekly,* February 25, 1933, 24–25.
Pinchot, Amos. "The Courage of the Cripple." *Masses* 9, no. 5 (March 1917): 19–21.
Pinkus, Karen. *Fuel: A Speculative Dictionary.* Minneapolis: University of Minnesota Press, 2016.
Porter, Theodore M. *Trust in Numbers: The Pursuit of Objectivity in Science and Public Life.* Princeton, NJ: Princeton University Press, 1995.
Pound, Ezra. *"Ezra Pound Speaking": Radio Speeches of World War II.* Edited by Leonard W. Doob. Westport, CT: Greenwood, 1978.
Preda, Roxana. "Social Credit in America: A View from Ezra Pound's Economic Correspondence, 1933–1940." *Paideuma* 34, no. 2/3 (2005): 201–231.
"Professor Ostwald and the University of Leipsic." *Outlook,* February 3, 1915, 284.
Purcell, Edward A. *The Crisis of Democratic Theory: Scientific Naturalism and the Problem of Value.* Lexington: University Press of Kentucky, 1973.
Rabinbach, Anson. *The Human Motor: Energy, Fatigue, and the Origins of Modernity.* New York: Basic Books, 1990.
Ramsey, P. W. "Technocracy, 'Energy Units' as a Basis for Exchange Fail to Give Proper Consideration to 'Human Element' Says Tech Economics Head." *Pittsburgh Post-Gazette,* December 21, 1932, 4.
Rand, Ayn. *The Fountainhead.* London: Penguin Books, 1994.
Rasmussen, Charles T., and Rick Tilman. *Jacques Loeb: His Science and Social Activism and Their Philosophical Foundations.* Philadelphia: American Philosophical Society, 1998.
Rasmussen, Charles T., and Rick Tilman. "Mechanistic Physiology and Institutional Economics: Jacques Loeb and Thorstein Veblen." *International Journal of Social Economics* 19, no. 10/11/12 (1992): 235–247.
Redman, Tim. *Ezra Pound and Italian Fascism.* Cambridge: Cambridge University Press, 1991.
Regal, Brian. *Henry Fairfield Osborn: Race, and the Search for the Origins of Man.* Aldershot: Ashgate, 2002.
Reich, Wilhelm. "Dialectical Materialism and Psychoanalysis" (1929). In *Sex-Pol,* edited by Lee Baxandall, 1–75. New York: Random House, 1966.

Reich, Wilhelm. *The Function of the Orgasm: Sex-Economic Problems of Biological Energy*. 1927. Reprint, New York: Farrar, Straus & Giroux, 1973.

Reich, Wilhelm. *The Mass Psychology of Fascism*. 1933. Reprint, New York: Orgone Institute Press, 1946.

Reingold, Nathan. "Jacques Loeb, The Scientist: His Papers and His Era." *Quarterly Journal of Current Acquisitions of the Library of Congress* 10 (1962): 119–130.

Ribuffo, Leo. "'Malaise' Revisited: Jimmy Carter and the Crisis of Confidence." In *The Liberal Persuasion: Arthur Schlesinger, Jr., and the Challenge of the American Past*, edited by John Patrick Diggins, 164–184. Princeton, NJ: Princeton University Press, 1997.

Rinkinen, Jenny, Elizabeth Shove, and Jacopo Torriti. *Energy Fables: Challenging the Ideas of the Energy Sector*. London: Routledge, 2019.

Rodgers, Daniel T. *Atlantic Crossings: Social Politics in a Progressive Age*. Cambridge, MA: Harvard University Press, 1998.

Roosevelt, Franklin D. "Campaign Address on Public Utilities and Development of Hydro-electric Power. Portland Oregon, September 21, 1932." In *The Public Papers and Addresses of Franklin D. Roosevelt*, vol. 1, *The Genesis of the New Deal*, 727–742. New York: Random House, 1938.

Roosevelt, Theodore. "The Law of Civilization and Decay." *Forum* 22 (January 1897): 575–589.

Roosevelt, Theodore. "Man and Statehood" (1901). In *The Strenuous Life: Essays and Addresses*, 113–120. Mineola, NY: Dover, 2009.

Roosevelt, Theodore. "The Origin and Evolution of Life." *Outlook*, January 16, 1918, 97–98.

Roosevelt, Theodore. "The Strenuous Life: Speech before the Hamilton Club, Chicago, April 10, 1899" (1899). In *The Strenuous Life: Essays and Addresses*, 1–10. Mineola, NY: Dover, 2009.

Rosenberg, Harold. "The American Action Painters" (1952). In *The Tradition of the New*, 23–69. New York: Horizon, 1959.

Ross, Dorothy. *The Origins of American Social Science*. Cambridge: Cambridge University Press, 1991.

Ross, Edward Alsworth. *Social Control: A Survey of the Foundations of Order*. London: Macmillan, 1901.

Rowland, Durbin W. "Man and Climate." *Ecology* 24, no. 4 (September 1, 1943): 505–506.

"Rubber on the Tennessee." *New York Times*, May 24, 1935, 20.

Rupp, Stephanie. "Considering Energy: $E = mc^2 = (magic.culture)^2$." In *Cultures of Energy: Power, Practices, Technologies of Energy*, edited by Sarah Strauss, Stephanie Rupp, and Thomas Love, 79–95. Walnut Creek, CA: Left Coast, 2013.

Samson, Paul R., and David Pitt. *The Biosphere and Noosphere Reader: Global Environment, Society and Change*. London: Routledge, 2002.

Samuel, Lawrence R. *Shrink: A Cultural History of Psychoanalysis in America*. Lincoln: University of Nebraska Press, 2013.

The Scenic Resources of the Tennessee Valley: A Descriptive and Pictorial Inventory. Prepared by the Department of Regional Planning Studies. Washington, DC: US Government Printing Office, 1938.

Schaffer, Daniel. "Benton MacKaye: The TVA Years." *Planning Perspectives* 5 (1990): 5–21.
Schaffer, Daniel. "Ideal and Reality in 1930s Regional Planning: The Case of the Tennessee Valley Authority." *Planning Perspectives* 1, no. 1 (1986): 27–44.
Scheibach, Michael. *Atomic Narratives and American Youth: Coming of Age with the Atom, 1945–1955*. Jefferson, NC: McFarland, 2003.
Schneider-Mayerson, Matthew. "Necrocracy in America: American Studies Begins to Address Fossil Fuels and Climate Change." *American Quarterly* 67, no. 2 (2015): 529–540.
Schulman, Bruce J. *The Seventies: The Great Shift in American Culture, Society, and Politics*. New York: Free Press, 2001.
Schuster, David G. *Neurasthenic Nation: America's Search for Health, Happiness, and Comfort, 1869–1920*. New Brunswick, NJ: Rutgers University Press, 2011.
"Scientist Declares World Is Held in Slavery by Sun." *Evening Sun* (Hanover, Pennsylvania), December 31, 1926, 8.
Sconce, Jeffrey. *Haunted Media: Electronic Presence from Telegraphy to Television*. Durham, NC: Duke University Press, 2000.
Sconce, Jeffrey. "Wireless Ego: The Pulp of Psychoanalysis." In *Broadcasting Modernism*, edited by Debra Rae Cohen, Michael Coyle, and Jane Lewty, 31–51. Gainesville: University Press of Florida, 2009.
Scott, Heidi. *Fuel: An Ecocritical History*. London: Bloomsbury Academic, 2018.
Scott, Howard. "Technology Smashes the Price System." *Harper's Monthly Magazine* 166 (January 1933): 129–142.
Scott, John, and Ray Bromley. *Envisioning Sociology: Victor Branford, Patrick Geddes, and the Quest for Social Reconstruction*. Albany: State University of New York Press, 2013.
Searle, G. R. *The Quest for National Efficiency: A Study in British Politics and Political Thought, 1899–1914*. Berkeley: University of California Press, 1971.
Secord, James A. "Knowledge in Transit." *Isis* 95 (2004): 654–672.
"Sees Christianity, Not Technocracy, as Cure." *Dayton Daily News*, January 17, 1933, 9.
Seow, Victor. *Carbon Technocracy: Energy Regimes in Modern East Asia*. Chicago: University of Chicago Press, 2022.
Shell, Marc. *Money, Language, and Thought: Literary and Philosophical Economies from the Medieval to the Modern Era*. Berkeley: University of California Press, 1982.
Sillen, Samuel. "Lewis Mumford's 'Mein Kampf.'" *New Masses*, October 15, 1940, 8–9.
Sluga, Glenda. *Internationalism in the Age of Nationalism*. Philadelphia: University of Pennsylvania Press, 2013.
Sluga, Glenda. *Nation, Psychology, and International Politics, 1870–1919*. New York: Palgrave Macmillan, 2006.
Sluga, Glenda. "UNESCO and the (One) World of Julian Huxley." *Journal of World History* 21, no. 3 (2010): 393–418.
Smil, Vaclav. *Energy in World History*. Boulder, CO: Westview, 1994.
Smith, Crosbie. "Natural Philosophy and Thermodynamics: William Thomson and 'The Dynamical Theory of Heat.'" *British Journal for the History of Science* 9, no. 3 (1976): 293–319.
Smith, Crosbie. *The Science of Energy: A Cultural History of Energy Physics in Victorian Britain*. Chicago: University of Chicago Press, 1998.

Smith, Crosbie, and M. Norton Wise. *Energy and Empire: A Biographical Study of Lord Kelvin*. Cambridge: Cambridge University Press, 1989.

Smith, Mark C. *Social Science in the Crucible: The American Debate over Objectivity and Purpose, 1918–1941*. Durham, NC: Duke University Press, 1994.

Smith, Roger. *Free Will and the Human Sciences in Britain, 1870–1910*. London: Pickering & Chatto, 2013.

Smith, T. L. "Notes and News." *American Journal of Psychology* 17, no. 1 (January 1906): 146.

Snow, C. P. *The Two Cultures*. Cambridge: Cambridge University Press, 1993.

Snyder, Carl. "Theory of Life: Dr. Jacques Loeb's Mechanistic Conception of Energy." *New York Times*, October 6, 1912, BR554.

Sobczynska, Danuta, and Ewa Czerwinska. "Wilhelm Ostwald and the German Monistic League on the Social and Cultural Role of Science." In *Wilhelm Ostwald at the Crossroads between Chemistry, Philosophy and Media Culture*, edited by Britta Görs, Nikos Psarros, and Paul Ziche, 59–68. Leipzig: Leipziger Universitätsverlag, 2005.

Soddy, Frederick. *Cartesian Economics: The Bearing of Physical Science upon State Stewardship*. London: Hendersons, 1922.

Soddy, Frederick. "Men, Machines, Money—and the Technocrats." *Daily Mail*, January 7, 1933.

Soddy, Frederick. *Wealth, Virtual Wealth and Debt: The Solution of the Economic Paradox*. New York: E. P. Dutton, 1933.

"Solar Bombardment." *Time* 35, no. 14 (April 1940): 12.

"Solar Storms Affect Reception More Than Weather Elements, According to Dr. Pickard—Sunspots Have Bad Influence on Ether." *New York Times*, January 16, 1927, X17.

Sorokin, Pitirim A. *Contemporary Sociological Theories*. New York: Harper & Brothers, 1928.

Sorokin, Pitirim A. "Is Accurate Social Planning Possible?" *American Sociological Review* 1, no. 1 (1936): 12–25.

Sorokin, Pitirim A. "A Neglected Factor of War." *American Sociological Review* 3, no. 4 (1938): 475–486.

Sorokin, Pitirim A. *Social and Cultural Dynamics: A Study of Change in Major Systems of Art, Truth, Ethics, Law, and Social Relationships*. Boston: Extending Horizons Books, 1957.

Spann, Edward K. *Designing Modern America: The Regional Planning Association of America and Its Members*. Columbus: Ohio State University Press, 1996.

Spann, Edward K. "Franklin Delano Roosevelt and the Regional Planning Association of America, 1931–1936." *New York History* 74, no. 2 (1993): 185–200.

Spengler, Oswald. *The Decline of the West*. Edited by Charles Francis Atkinson. New York: A. A. Knopf, 1926.

Spindler, Michael. *Veblen and Modern America: Revolutionary Iconoclast*. London: Pluto, 2002.

"Sport Our Colleges' Bane: Prof. Ostwald Says It Draws Attention from Intellectual Pursuits." *New York Times*, February 21, 1906, 1.

"Stature of Man Found Decreasing in Tests Made on 65,0000 Students." *New York Times*, May 4, 1940, 19.

Stein, Clarence S. "Dinosaur Cities." *Survey Graphic* 54, no. 3 (May 1, 1925): 134–138.
Stetson, Harlan T. *Earth, Radio and the Stars*. New York: Whittlesey House, McGraw-Hill, 1934.
Stetson, Harlan T. *Man and the Stars*. New York: Whittlesey House, McGraw-Hill, 1930.
Stetson, Harlan T. *Sunspots and Their Effects*. New York: Whittlesey House, McGraw-Hill, 1937.
Stevens, Jay. *Storming Heaven: LSD and the American Dream*. New York: Atlantic Monthly Press, 1987.
Stewart, Janet. "Sociology, Culture and Energy: The Case of Wilhelm Ostwald's 'Sociological Energetics'—a Translation and Exposition of a Classic Text." *Cultural Sociology* 8, no. 3 (2014): 333–350.
Stoekl, Allan. *Bataille's Peak: Energy, Religion, and Postsustainability*. Minneapolis: University of Minnesota Press, 2007.
Stokley, James. "Radio Reception and Sun Spots." *Science News-Letter* 16, no. 434 (1929): 59–61.
Storck, John. "The Dynamics of Our Society: *The God of the Machine*, by Isabel Paterson." *New York Times*, May 23, 1943, BR29.
Strick, James Edgar. *Wilhelm Reich, Biologist*. Cambridge, MA: Harvard University Press, 2015.
Sulloway, Frank J. *Freud, Biologist of the Mind: Beyond the Psychoanalytic Legend*. Cambridge, MA: Harvard University Press, 1979.
"A Summons to Save the Wilderness." *Living Wilderness* 1 (September 1935): 1.
"Sunspots and Sudden Deaths Associated by French Doctor." *New York Times*, March 2, 1927, 1.
"Sunspots and Terrestrial Phenomena." *Evening Telegraph*, June 4, 1902, 4.
"Sun-Spots as Prosperity Regulator." *Literary Digest* 122, no. 4 (July 25, 1936): 16–17.
"Sunspots Called Key to the Future; Russian Professor Finds Parallel between Human and Solar Activities and Would Time Events Accordingly." *New York Times*, January 30, 1927, xx9.
"Sunspots Cause of Wars?" *Popular Science*, June 1927, 50.
"Sunspots Down." *Time* 32, no. 2 (November 7, 1938): 24.
"Sunspots Forecast Better Radio." *Popular Mechanics* 50, no. 4 (October 1928): 594–597.
"Sunspots Grow, and Studies Suggest Effect Human Conduct." *Evening News, Tonawanda*, February 24, 1938, 7.
Surén, Hans. *Man and Sunlight*. Translated by David Arthur Jones and C. W. Saleeby. Slough: Sollux, 1927.
Surette, Leon. *Pound in Purgatory: From Economic Radicalism to Anti-Semitism*. Champaign: University of Illinois Press, 1999.
Sutter, Paul. *Driven Wild: How the Fight against Automobiles Launched the Modern Wilderness Movement*. Seattle: University of Washington Press, 2002.
Sutter, Paul. "Putting the Intellectual Back in Environmental History." *Modern Intellectual History* 18, no. 2 (2020): 596–605.
Sutter, Paul. "'A Retreat from Profit': Colonization, the Appalachian Trail, and the Social Roots of Benton MacKaye's Wilderness Advocacy." *Environmental History* 4, no. 4 (1999): 553–577.

Swer, Gregory Morgan. "Technics and (Para)praxis: The Freudian Dimensions of Lewis Mumford's Theories of Technology." *History of Human Sciences* 17, no. 4 (2004): 45–68.

Swetlitz, Marc. "Julian Huxley and the End of Evolution." *Journal of the History of Biology* 28, no. 2 (July 1, 1995): 181–217.

Szeman, Imre, and Dominic Boyer. *Energy Humanities: An Anthology*. Baltimore: Johns Hopkins University Press, 2017.

Szeman, Imre, Jennifer Wenzel, and Patricia Yaeger. *Fueling Culture: 101 Words for Energy and Environment*. New York: Fordham University Press, 2017.

Tansley, Arthur. *The New Psychology and Its Relation to Life*. London: George Allen & Unwin, 1920.

Tattersall, Mason. "Thermal Degeneration: Thermodynamics and the Heat-Death of the Universe in Victorian Science, Philosophy, and Culture." In *Decadence, Degeneration and the End: Studies in the European Fin de Siècle*, edited by Marja Härmänmaa and Christopher Nissen, 17–34. New York: Palgrave Macmillan, 2014.

Tchijevsky, Alexander. "Physical Factors of the Historical Process" (1926). Translated by V. P. De Smitt. *Cycles*, January 1971, 11–27.

Teale, Edwin. "New Discoveries Show Electricity Governs Our Lives." *Popular Science Monthly* 124, no. 2 (February 1934): 11–13, 100–102.

"Technocracy Idea Borrowed, He Says." *New York Times*, December 30, 1932, 13.

"Technocracy Like Dream of Ostrich, House Told." *Los Angeles Times*, January 15, 1933, 10.

"Technocrat Gives 'Energy' as a Fee." *New York Times*, January 9, 1933, 28.

Teilhard de Chardin, Pierre. *The Phenomenon of Man*. With an introduction by Julian Huxley. New York: Harper, 1959.

Teller, C. "Man as Mechanism: The Extraordinary Experiments of Jacques Loeb." *Metropolitan Magazine*, November 1912, 63.

"That Rubber Yard." *New York Times*, November 4, 1934, 4.

Thomas, John. L. "The Uses of Catastrophism: Lewis Mumford, Vernon L. Parrington, Van Wyck Brooks, and the End of American Regionalism." *American Quarterly* 42, no. 2 (1990): 223–251.

Thomson, William. "On a Universal Tendency in Nature to the Dissipation of Mechanical Energy." *Philosophy Magazine* 4 (1852): 304–306.

"Those Sinister Spots." *Pittsburgh Press*, October 9, 1937, 4.

Tichi, Cecelia. *Shifting Gears: Technology, Literature, Culture in Modernist America*. Chapel Hill: University of North Carolina Press, 1987.

Titus, E. K. "Freak Weather Seen Sign of Grave Climatic Changes." *Brooklyn Daily Eagle*, January 8, 1928, B3.

Tobey, Ronald C. *Technology as Freedom: The New Deal and the Electrical Modernization of the American Home*. Berkeley: University of California Press, 1996.

Toews, John. E. "Intellectual History after the Linguistic Turn: The Autonomy of Meaning and the Irreducibility of Experience." *American History Review* 92, no. 4 (1987): 879–907.

Topham, Jonathan R. "Focus: Historicizing 'Popular Science.'" *Isis* 100, no. 2 (2009): 310–318.

Townsend, Kim. *Manhood at Harvard: William James and Others*. New York: W. W. Norton, 1996.
Toynbee, Arnold. "Does History Always Repeat Itself? Professor Toynbee, Examining the Rise and Fall of Civilizations, Finds Hope for Ours." *New York Times*, September 21, 1947, SM15.
Toynbee, Arnold. *A Study of History*. Edited by D. C. Somervell. Oxford: Oxford University Press, 1947.
Trentmann, Frank. "Civilization and Its Discontents: English Neo-Romanticism and the Transformation of Anti-modernism in Twentieth-Century Western Culture." *Journal of Contemporary History* 29, no. 4 (October 1, 1994): 583–625.
Trentmann, Frank, and Anna Carlsson-Hyslop. "The Evolution of Energy Demand in Britain: Politics, Daily Life, and Public Housing, 1920s–1970s." *Historical Journal* 61, no. 3 (2018): 807–839.
"True Winters Are Not What They Used to Be." *New York Times*, January 22, 1939, 98.
Trump, Donald. "Remarks by President Trump at the Unleashing American Energy Event." Speech, US Department of Energy, Washington, DC, June 29, 2017. https://web.archive.org/web/20210120195650/https://trumpwhitehouse.archives.gov/briefings-statements/remarks-president-trump-unleashing-american-energy-event/.
Turnbull, Thomas. "From Incommensurability to Ubiquity: An Energy History of Geographic Thought." *Journal of Historical Geography* 73 (2021): 11–23.
Turner, Christopher. *Adventures in the Orgasmatron*. London: Harper, 2008.
Tyler, J. Ross. *Financial Astrology*. Chicago: D. G. Nelson, 1947.
Tyndall, John. *Address Delivered before the British Association Assembled at Belfast, with Additions, 1874*. London: Longmans, Green, 1874.
Tyrrell, Ian R. *Crisis of the Wasteful Nation: Empire and Conservation in Theodore Roosevelt's America*. Chicago: University of Chicago Press, 2015.
Underwood, Ted. *The Work of the Sun: Literature, Science, and Political Economy, 1760–1860*. New York: Palgrave Macmillan, 2005.
United World Federalists (US). *One World . . . or None*. Cleveland: United World Federalists, 1957.
Unwin, J. D. *Hopousia; or, The Sexual and Economic Foundations of a New Society*. London: George Allen & Unwin, 1940.
Unwin, J. D. *Sex and Culture*. Oxford: Oxford University Press, 1934.
Veblen, Thorstein. *The Engineers and the Price System*. 1921. Reprint, New York: Harcourt, Brace & World, 1963.
Veblen, Thorstein. *The Instinct of Workmanship, and the State of the Industrial Arts*. New York: Macmillan, 1914.
Veblen, Thorstein. *The Theory of the Leisure Class*. 1899. Reprint, Boston: Houghton Mifflin, 1973.
Veder, Robin. *The Living Line: Modern Art and the Economy of Energy*. Hanover, NH: Dartmouth College Press, 2015.
Vernadsky, Vladimir, and Mark A. McMenamin. *The Biosphere*. Translated by David B. Langmuir. 1926. Reprint, New York: Copernicus, 1998.
"Vox Populi Hails Technocracy—or Anything Else That'll Help." *Brooklyn Daily Eagle*, January 17, 1933, 3, 20.

"Warm Climate Thought Cause of Youth Becoming Shorter." *Los Angeles Times*, May 8, 1940, 20.

"Warmer World." *Time* 33, no. 1 (January 2, 1939): 33.

Warren, Donald I. *Radio Priest: Charles Coughlin, the Father of Hate Radio*. New York: Free Press, 1996.

Watts, Sarah Lyons. *Rough Rider in the White House: Theodore Roosevelt and the Politics of Desire*. Chicago: University of Chicago Press, 2003.

Weart, Spencer R. *The Discovery of Global Warming*. Cambridge, MA: Harvard University Press, 2003.

"The Weather and Our Feelings." *Forum* 74, no. 3 (September 1925): 321–332.

Webb, George Ernest. *Tree Rings and Telescopes: The Scientific Career of A. E. Douglass*. Tucson: University of Arizona Press, 1983.

Weir, Todd H. *Monism Science, Philosophy, Religion, and the History of a Worldview*. New York: Palgrave Macmillan, 2012.

Wells, H. G. "Ideals of Organization." *New Republic* 3, no. 38 (July 24, 1915): 301–302.

Wells, H. G. *A Modern Utopia*. Edited by Gregory Claeys and Patrick Parrinder. 1905. Reprint, London: Penguin Books, 2005.

Wells, H. G. *The Time Machine*. Edited by Patrick Parrinder. 1895. Reprint, London: Penguin Books, 2005.

Wells, H. G., Julian Huxley, and G. P. Wells. *The Science of Life*. Garden City, NY: Doubleday, Doran, 1931.

Wellum, Caleb. *Energizing Neoliberalism: The 1970s Energy Crisis and the Making of Modern America*. Baltimore: Johns Hopkins University Press, 2023.

Wellum, Caleb. "The Use of Energy History." *Modern American History* 6 (2023): 201–219.

Welter, Volker. *Biopolis: Patrick Geddes and the City of Life*. Cambridge, MA: MIT Press, 2002.

Westbrook, Robert B. *John Dewey and American Democracy*. Ithaca, NY: Cornell University Press, 1991.

"What Price Energy?" *Newsweek*, May 2, 1977, 12.

Wheatland, Thomas. *The Frankfurt School in Exile*. Minneapolis: University of Minnesota Press, 2009.

Wheeler, Raymond Holder. "The Effect of Climate on Human Behavior in History." *Transactions of the Kansas Academy of Science* 46 (1943): 33–51.

Wheeler, Raymond Holder. *The Laws of Human Nature*. London: Nisbet, 1931.

Wheeler, Raymond Holder, and Michael G. Zahorchak. *Climate, the Key to Understanding Business Cycles, with a Forecast of Trends into the 21st Century*. Linden, NJ: Tide, 1980.

Wiener, Norbert. *Cybernetics, or Control and Communication in the Animal and the Machine*. 1948. Reprint, Cambridge, MA: MIT Press, 1965.

"The Wilderness Society Platform." *Living Wilderness* 1 (September 1935): 2.

Wilkinson, W. E. Ayton. "Will-Force and the Conservation of Energy." *Monist* 18, no. 1 (January 1, 1908): 1–20.

Williams, William A. "Brooks Adams and American Expansion." *New England Quarterly* 25, no. 2 (June 1952): 217–232.

Wilson, Colin. *The Quest for Wilhelm Reich*. Garden City, NY: Anchor Press / Doubleday, 1981.

Wilson, Guy, Dianne H. Pilgrim, and Dickran Tashjian Wilson. *The Machine Age in America, 1918–1941*. New York: Brooklyn Museum, 1986.
Wilson, Sheena, Adam Carlson, and Imre Szeman. *Petrocultures: Oil, Politics, and Culture*. Montreal: McGill–Queen's University Press, 2017.
Wilson, Woodrow. *The New Freedom: A Call for the Emancipation of the Generous Energies of a People*. New York: Doubleday & Page, 1913.
Woiak, Joanne. "Designing a Brave New World: Eugenics, Politics, and Fiction." *Public Historian* 29, no. 3 (August 1, 2007): 105–129.
"World Climate Growing Warmer, Says Russians, Citing Arctic Data." *New York Times*, December 12, 1938, 1.
Worpole, Ken. *Here Comes the Sun: Architecture and Public Space in Twentieth-Century European Culture*. London: Reaktion, 2000.
Wright, Rebecca. "Sunspots and Sync." In *Energy and Literature: Essays on Energy and Its Social and Environmental Implications in Twentieth and Twenty-First Century Literary Texts*, edited by Paula Anca Farca, 9–19. Oxford: TrueHeart, 2015.
Wright, Rebecca, Hiroki Shin, and Frank Trentmann. *Power, Energy and International Cooperation: A History of the World Energy Council*. Munich: Oekom, 2019.
Wright, Rebecca, and Frank Trentmann. "The Social Life of Energy Futures: Experts, Consumers and Demand in the Golden Age of Modernization, c. 1900–1973." In *Economy and Environment in the Hand of Experts*, edited by Frank Trentmann, Anna Barbara Sum, and Manuel Rivera, 47–77. Munich: Oekom, 2018.
Wrigley, E. A. *Energy and the English Industrial Revolution*. Cambridge: Cambridge University Press, 2010.
Yaeger, Patricia. "Editor's Column: Literature in the Ages of Wood, Tallow, Coal, Whale Oil, Gasoline, Atomic Power, and Other Energy Sources." *PMLA* 126, no. 2 (2011): 305–326.
"A Yardstick No More." *New York Times*, July 14, 1935, 8.
Yglesias, Matthew. "'Low-Energy': Donald Trump's Favorite Diss on Jeb Bush, Explained." *Vox*, September 4, 2015.
Young, George M. *The Russian Cosmists: The Esoteric Futurism of Nikolai Fedorov and His Followers*. Oxford: Oxford University Press, 2012.
Ziser, Michael, Natasha Zaretsky, and Julie Sze. "Perpetual Motion: Energy and American Studies." *American Quarterly* 72, no. 3 (2020): 543–557.

INDEX

Aberhart, William, 119
A+B theorem, 118–19, 120
abstract expressionism, 149–50
Acker, Antoine, 8
action painting, 149
Adams, Brooks: Huntington comparison, 68; Law, 43–44, 120, 172; on Progressive Party, 50; and will, 26, 41–44, 49; and World War I, 51
Adams, Henry, 26, 41, 42–43, 44–45, 132
Addams, Jane, 75
Adorno, Theodor, 153
Alpha and Beta marriage, 160
American Eugenics Society, 73
Americanness and American culture: American energy as threatened, 3, 17–18, 25, 26, 56–57, 71, 73, 74, 77–79, 172; American exceptionalism, 73, 79; and climatic energy, 17–18, 56–57, 71, 73, 74, 77–79, 172; and energy (power) abundance, 6–7; and energy (power) crisis, 171–72, 174; and heroic individual, 145–50; measuring of American body, 56–57; and misdirected energy, 57; and pop culture, 153; and primordial energy, 90, 100–102; and psychic energy, 83–84, 85–87; and wilderness, 100–102
Amory, D. C. Copley, 58
animals and sunspots, 58, 60
Anthropocene, 8–9
anti-Semitism, 42, 109, 120–22
Appalachian Trail, 91, 97
art: clash of religion, 145; and heroic individual / new man concept, 19, 142, 149–50
atomic bomb, anxiety over, 139–41

authoritarianism: and energetic ethics, 19, 154, 156; and Gardiner, 91; and Julian Huxley, 163–64; and replacing money with energy, 115–16; and sex, 160. *See also* totalitarianism

bathmic energy / bathmism, 39
Baumes Committee, 75
Beard, George Miller, 40, 41, 73, 196n119
Becker, Valeska, 126
Bergson, Henri, 12, 29, 36, 37, 165
Bing, Alexander, 85
Black, Brian, 4
Bliven, Bruce, 144
blood-consciousness, 90
body: climate's effect on, 56–57, 71, 78; as machine/motor, 6, 11; measuring of American, 56–57; and race, 71
Boltzmann, Ludwig, 32
Bourne, Randolph, 1–2
Bourne-Dewey debate, 1–2, 6
Boyer, Dominic, 4, 5, 141
Brahman principles, 157, 158
Bridge, The (Die Brücke), 31–32
Bruckner cycle, 55
Bruère, Robert, 96
Buell, Frederick, 4
Bunsen-Roscoe law of reciprocity, 27

Caldwell, Orestes, 60
Campbell, Joseph, 148
capitalism: and authoritarianism, 154; critiques of, 86, 89, 114, 145; and energy metaphors, 174; and fossil fuels, 4; laissez-faire, 131, 132, 139, 141, 173; and TVA, 104, 126; and will, 24, 37–38, 43, 47

Carnegie, Andrew, 49
Carter, Jimmy, 4, 171–72, 174, 175
Carus, Paul, 32, 33
Chase, Richard, 156–57
Chase, Stuart: and communism, 124; as Hell Raiser, 97; on New Deal, 124–25; and Paterson, 131; and RPAA, 85; and Soviet Union, 95, 151; on wasted energy, 108, 125, 173
Chatterjee, Elizabeth, 8
children and youth, 49, 74–76
China: Chinese ideograms, 122; energy consumption by, 2
cities: and critiques of modernity, 88; and energy (power), 94–95, 173; energy basis of, 125–26; migration away from, 94–99, 100
civilization: Brooks Adams's Law of, 41–42, 120, 172; and climatic energy, 68–73, 77–79; decline/collapse fears, 72–73, 77–79, 88, 146–48; and energetic ethics, 144–48; and energy basis of cities, 125–26; and etherialization, 147–48; and heroic individual, 146, 148; Huntington's measures of, 69; and libido, 85–87; and new man concept, 144–45; and psychic energy, 87–91; and race, 167; and sex, 158–60; shifts in centers of, 71–72; and societies of status/contract, 133; and Spengler, 88, 98; and wilderness, 100; and will, 146
Civilization and Its Discontents (Freud), 86–87
class: and neurasthenia, 41; and predatory culture, 36; and replacing money with energy, 150–51
climate change, 64, 78–79, 174
climatic determinism, 17–18, 56, 57, 68–73, 79
climatic energy: and civilization, 68–73, 77–79; and control of energy, 56–57, 74–77; and culture, 64–66, 68–73; and energy language, 17–18, 173; and energy paradox, 57, 80; and human behavior, 55–57, 60–68; and misdirected energy, 57, 74–77, 80, 172; overview of, 17–18, 55–57; and race, 17–18, 56–57, 65, 70–71, 73, 76, 78–79; and sunspots, 56, 57–68, 78; as threat to America, 17–18, 56–57, 172
cloud-buster, 143
collectivism: and monism, 32; and Paterson, 131, 132; and Rand, 152–53; in Soviet Union, 95, 152; and will, 17, 35

colonialism and climatic energy, 71
Columbia University, 108, 113, 114, 153
Committee on Technocracy, 108, 113, 114
Committee on the Atmosphere of Man, 69
communism: and Chase, 124; decline in Western support for, 141; and Paterson, 131, 133; and Reich, 81, 86; in Soviet Union, 95, 152; and TVA, 104
conservation, laws of. *See* thermodynamics, laws of
conservation movement (ecology), 23, 100–102
conservation of energy (power), 4, 171–72, 174
consumerism and consumption: and energy crisis in 1970s, 171; and totalitarianism, 142; and waste of energy, 36
contract, societies of, 133
control, energy as in need of: and atomic age, 139–42; and climatic energy, 56–57, 74–77; and efficiency movement, 48; and energetic ethics, 139–42; and individualism, 143, 164–65; in overview, 7–8; and state control, 13–14
Cooke, George Willis, 37
Cooke, Morris L., 94
Coolidge, Calvin, 61
Coolidge, Grace, 126
Cope, Edward Drinker, 39
cosmecology, 59–60
Coughlin, Charles (Father), 120–21
Cousins, Norman, 139–40, 162
Cowley, Malcolm, 144
Crick, Walter, 120
crime, 74–76
Cronon, William, 9
Crutzen, Paul, 8
culture: and climatic energy, 64–66, 68–73; energy's (power) effect on culture, 5, 25, 82–84, 95; Faustian, 88; and heroic individual, 145–50; and internationalization, 166; and libido, 86–87; and orgone, 81; pop culture, 153; predatory, 36–37; and psychic energy, 18, 81–87, 97–99; and sunspots, 64–66; and TVA, 102–5; and wilderness, 100–102. *See also* Americanness and American culture

Daggett, Cara New, 4, 5–6, 7, 174, 175
dance, 89, 90
Danube Valley Authority, 105

Davis, Cushman, 43
deadly orgone, 143
death drive (Thanatos), 12, 86
Deerfield Water Power Project, 96
de la Peña, Carolyn, 4, 25, 83
democracy: and climatic energy, 61, 65; and energetic ethics, 139, 141–42, 143, 151, 154, 156, 162, 164; and Julian Huxley, 16, 162, 164; and monism, 30; and psychic energy, 81, 86, 88, 90; and rationality of democratic subject, 12; and Reich, 81, 86; and replacing money with energy, 115; and Spengler, 88; and Teddy Roosevelt, 47, 51; and TVA, 104, 162; and US intervention in World War I, 1, 2, 6; and will, 17, 25, 43–44, 50–51
demon, Maxwell's, 33–34
determinism: climatic determinism, 17–18, 56, 57, 68–73, 79; and monism, 33–34
devolutionists, 40–44
Dewey, John, 1–2, 11, 151–52
dictatorship. *See* authoritarianism; totalitarianism
Director, Aaron, 112, 114–15
dog, heliotropic, 27, 31
dor (deadly orgone), 143
Douglas, C. H., 109–10, 112, 118, 120
Douglass, Andrew Ellicott, 58, 78
Draper, Earle, 102
Dreiser, Theodore, 37
Driesch, Hans, 29
drugs, 158, 168–69
dynamogenic will, 46–47

economics: accounting and objective value of energy, 10; and solar credit, 118–19; and Technocracy, 11, 18, 107, 109–18. *See also* money, replacing with energy
Edge, Charles N., 107
education, 166
efficiency: and climate's effect on body, 56; efficiency movement, 15, 48; Germany as over-worshipping, 12, 49–50, 51; industrial versus social in MacKaye, 98; and monism, 30
ego, 87, 155, 199n27
Einstein, Albert, 129, 130, 140, 165
Eisenhower, Dwight, 4
élan vital, 29, 37

electricity: and culture, 83–84, 95; and Giant Power, 82, 84, 94–99; scholarship focus on, 4; and Soviet Union, 95; spread of in United States, 25; and TVA, 18, 83–84, 102–5, 123, 161–62
Elton, Charles, 58
End Poverty in California movement, 119
energetic ethics: and Aldous Huxley, 19, 139, 151, 156–61; and atomic bomb anxiety, 139–41; and democracy, 139, 141–42, 143, 151, 154, 156, 162, 164; and energy paradox, 155–61; and heroic individual, 19, 139, 142, 143–50, 153, 164–65, 168; and internationalism, 19, 165–68, 169; and Julian Huxley, 19, 139–40, 143, 161–65, 173; and new man concept, 142, 143–45, 168; overview of, 19, 139–43; and politics, 19, 139, 142–43, 161–65; and totalitarianism, 19, 141–42, 150–54, 156, 164, 168; and will, 139, 140, 146
energetic geometry, 129
energetics, 15, 30–35, 36, 67–68, 109
energy: breadth of contexts in discourse, 6, 14–16; versus force, 1–2; mechanistic concept of, 25, 26–30, 35–37, 51; as term, 5; vitalist concept of, 25, 29–30. *See also* climatic energy; control, energy as in need of; good, energy as innately; intrinsic/objective value, energy as; money, replacing with energy; paradox, energy; primitive/primordial energy; psychic energy; relative value, energy as; yardstick, energy as
energy (power). *See* power (energy)
energy certificates, 113–14, 117, 122, 126–27, 134–35
energy consciousness, development of, 2–4, 172, 175
energy consumption: in China, 2; in United States, 2, 6–7, 8, 25
energy fables, 174–75
energy humanities, overview of scholarship, 4–10
energy language: and body, 6; and climatic energy, 17–18, 173; development of, 5–6; and energy as term, 5; and gender, 16; and infrastructure planning, 83–84, 173; and monism, 15; and politics, 3–4, 18, 109–10, 122–26, 134, 173; and psychology, 83; and race, 16; and work, 6, 7, 174–75
Energy Survey of North America, 107–8, 113

entelechy, 29, 39
eotechnic phase of civilization, 83
erg, replacing money with, 107, 119
Erginette, 107, 108, 117, 126, 135
Eros, 12, 86
Erskine, John, 107, 126
etherialization, 147–48
eugenics, 17, 56, 57, 73
evolution: devolutionists, 40–44; and psychic energy, 167; social evolution phases in Mumford, 83; and will, 37–44
evolutionary humanism. *See* scientific humanism
excitability, index of mass human, 61

fables, energy, 174–75
fascism: and anti-Semitism, 120, 122; and Mumford, 150–51, 156; and Paterson, 131, 132, 133; and Pound, 18, 121; and primordial energy movements, 91; and Spengler, 88, 89; as threat, 104, 120, 142, 144, 150–51, 156
Faustian culture, 88
fear and civilization, 41–42, 43
Fischer, Louis, 152
Fisher, Irving, 113
folk dance, 89, 90
force: versus energy in James, 45; persistence of force model and Spencer, 45; and US involvement in World War I, 1–2
fossil fuels: and Anthropocene, 8–9; and energy consciousness development, 3; versus energy discourse, 5, 174; and moral failure, 174; scholarship focus on, 4–6, 7, 9
Fountainhead, The (Rand), 152–53
Frankfurt School, 19, 142, 153, 154
Franzese, Sergio, 48
Frazer, Harold, 134
Frederick, Christine, 117
free will. *See* will
Freud, Sigmund, 12, 18, 85, 86, 92, 154
Fromm, Erich, 153–55
frontier thesis, 41, 100
Fuller, Buckminster, 110, 128–31

Gardiner, Rolf, 90–91
Geddes, Patrick, 16, 93, 114, 198n7
gender: and energy language, 16; and replacing money with energy, 117; and waste of energy, 207n75. *See also* masculinity; women

geographic and waterway metaphors for psychic energy, 91–94, 99, 104–6
Geographic School, 67–68
geometry, energetic, 129
German Monist League, 15, 27, 30, 89
Germany: folk and youth movements, 90–91; hyperinflation in, 113; and National Socialism, 89, 185n40; as over-worshipping energy and efficiency, 12, 49–50, 51
Gesell, Silvio, 121–22
Giant Power, 82, 84, 94–99
Gilbreth, Frank, 130
Gilbreth, Lillian, 130
Gilfillan, S. Colum, 71–72
GOELRO (State Commission for the Electrification of Russia), 95
Gold, Barri J., 5
Golden Day, The (Mumford), 84–85
gold standard, 113, 129
Gompers, Samuel, 49
good, energy as innately: in overview, 7–8; social good and climatic energy, 57; social good and intrinsic value of energy, 10; and Technocracy critiques, 131–32
Goodman, Paul, 155–56
Great Depression: and New Deal, 18, 109–10, 120–26, 131, 132, 133, 134, 140, 161; and Technocracy, 107–18
greed and civilization, 41–42, 43, 44, 51
Green New Deal, 4
Green Shirts, 118–19
Griffiss, Barton, 126
Gross, A. O., 58

Haeckel, Ernst, 30, 89
Hale, George Ellery, 58
Hall, G. Stanley, 49
Halliday, Sam, 4
hallucinogenics, 158, 168–69
Hammond, John Hays, Jr., 27
happiness, 15, 187n77
Hargrave, John, 118–20
Hayek, Friedrich, 128
Hazlitt, Henry, 116–17, 127–28
Heard, Gerald, 156–57
heliotherapy, 89
heliotropism, 26–27, 28, 31, 42
Heller, Harriet, 75
Hell Raiser, 97

Helmholtz, Hermann von, 10, 33
Henriksen, Margot, 139
heroic individual: and energetic ethics, 19, 139, 142, 143–50, 153, 164–65, 168; hero archetype, 148, 149; and new man concept, 142, 143–45, 168; and strenuous life, 47; and will, 46, 47
Hippocrates, 68
Hoke, Travis, 79
Hook, Sidney, 151
Hoover, Herbert, 115
Hoover, Lou Henry, 126
Horkheimer, Max, 153
Howard, Ebenezer, 16
Huber, Matthew T., 4
human entropy, 159–60
humanism, scientific, 162–65, 166–68, 169
Huntington, Ellsworth: and climatic determinism, 17, 56, 57, 68–73, 79; and decline of civilization, 77; influence on Toynbee, 214n45; measures of civilization, 69; and sunspots, 58, 67, 68–73
Huxley, Aldous, 19, 139, 151, 156–61, 168
Huxley, Julian, 19; background, 140, 161; and drugs, 169; and energetic ethics, 19, 139–40, 143, 161–65, 173; influence of, 16, 157; and psychoanalysis, 161, 162, 166, 173; and scientific humanism, 162–65, 166–68, 169; on TVA, 104, 105, 161–62; and UNESCO, 19, 139, 143, 162, 165–68, 169
Huxley, Thomas Henry, 99, 140

id, 87, 199n27
ideal liberalism, 144, 152
ideograms, 122
idolum, 145
immigration: and climatic energy, 16, 17, 56, 57, 73, 76–77, 79–80; and misdirected energy, 76–77, 172
imperialism: and climatic energy, 71; Julian Huxley on, 160, 167; US, 43, 47–48, 71
index of mass human excitability, 61
individualism: and climatic energy, 65, 67; and energetic ethics, 19, 139, 142, 143–50, 153, 164–65; and heroic individual, 19, 46, 47, 139, 142, 143–50, 153, 168; and infrastructure planning, 104; and new man concept, 142, 143–45, 168; and Ostwald, 32, 47, 49; and Rand, 153; and Teddy Roosevelt, 47–48; and wealth, 35, 36; and will, 17, 25, 34, 35–36, 44, 46, 47–48, 49, 50–51
information theory, 173–74
infrastructure planning: and energetic ethics, 161; and energy language, 83–84, 173; and energy paradox, 18, 84, 106; and Giant Power, 82, 84, 94–99; and internationalism, 105, 161; and totalitarianism, 141; and TVA, 18, 83–84, 102–5, 123, 161–62. *See also* Regional Planning Association of America (RPAA)
internationalism and world unity: and energetic ethics, 19, 139–40, 142, 165–68, 169; and infrastructure planning, 105, 161; and Julian Huxley, 162–65, 166–68, 173; and MacKaye, 105, 106; and Mumford, 162; and race, 167; and replacing money with energy, 165; and scientific humanism, 162–65, 166–68, 169; travel and correspondence, 15–16; and will, 139–40
intrinsic/objective value, energy as: climatic energy, 57; as debatable, 7, 10–14, 17–19, 173, 175; and energy as yardstick, 11, 109; and replacing money, 109–10, 113–14, 116–17, 118, 123, 130, 131–34, 135; and social good, 10; and US involvement in World War I debate, 1–2, 12; and will, 26. *See also* relative value, energy as

James, William: and energy language, 3, 4; and heroic individual, 46, 47; and quality versus quantity of energy, 13; and relative value of energy, 13, 14; and revelation, 163; and waterway metaphors for psyche, 92; and will, 17, 26, 44–48, 49, 51–52. *See also* moral equivalent of war concept
Jevons, William Stanley, 57–58
Johnson, Bob, 3, 4, 7, 25, 83
Johnson, Christopher, 9
Johnson, Gerald W., 109
Jordan, John M., 11
Jordan, Virgil, 109
Jung, Carl, 148, 201n69
juvenile delinquency, 75

Kennedy, John F., 3–4
Kern, Stephen, 5, 25
Keun, Odette, 104
Kindred of the Kibbo Kift, 89–90, 91, 118
Kitson, Arthur, 120

Labanca, Nicola, 174
laissez-faire capitalism, 131, 132, 139, 141, 173
Lawrence, D. H., 84, 90, 91
LeMenager, Stephanie, 4
lemmings, 58
Lenin, Vladimir, 95
Leopold, Aldo, 58, 100, 101
Lewis, Sinclair, 37, 119
liberalism, ideal versus pragmatic, 144, 150–51
libido, 12, 81, 84–87, 92, 93–94, 145, 154
Lieberman, Jennifer, 4, 83
"Life Force" concept, 29, 37
light and heliotropism, 26–27, 28, 30, 42
Lilienthal, David, 103, 104, 124
Lippmann, Walter, 13–14, 115, 127
Lodge, Henry Cabot, 43, 51
Loeb, Harold, 115, 116
Loeb, Jacques, 16, 17, 23–30, 35–37, 42, 51
Lutz, Tom, 25

MacDonald, Dwight, 141, 155
MacDuffie, Allen, 5
Mach, Ernst, 34, 35–36
MacKaye, Benton, 18, 84, 85, 88, 91, 96–105, 106
MacLeish, Archibald, 115, 151
Malm, Andreas, 8
manifest destiny, 40
Marsh, George Perkins, 48
Marshall, Robert "Bob," 101
Martin, Edward S., 60, 61, 63–64
Martinez-Alier, Joan, 109, 113
masculinity: and energy language, 16; and Teddy Roosevelt, 24, 47
mass man, 144, 154
Matamek Conference on Biological Cycles, 58, 60
Maxwell, James Clerk, 33–34
Maxwell's demon, 33–34
McCrae, John, 110–11
mechanistic concept of energy, 25, 26–30, 35–37, 51
Mills, Clarence Alonzo, 74, 76–79
Mills, C. Wright, 155–56
Mirowski, Philip, 10, 109
misdirected energy, 57, 74–77, 80, 85, 172
Mitchell, Timothy, 4
modernity: and neurasthenia, 41; psychic energy and critiques of, 84, 87–91

money, as form of energy, 41–42
money, replacing with energy: and anti-Semitism, 120–22; critiques of, 126–28; and energy certificates, 113–14, 117, 122, 126–27, 134–35; and energy paradox, 110, 134, 135; and erg, 107, 119; and Fuller, 110, 128–31; and gold standard, 113, 129; and internationalism, 165; and laws of thermodynamics, 110–12, 120, 121; and objectivism, 18; in overview, 11, 18, 107–10; and Paterson, 18, 110, 131–35; and politics, 18, 109–10, 112, 115, 118–22; and Social Credit movement, 18, 109–10, 112, 118–20, 121; and Technocracy, 11, 18, 107, 109–18, 126–28; and therblig standard, 130
monism: and critique of modernity, 89; and Loeb, 17, 26, 27, 35; and Ostwald, 15, 17; and will, 17, 25–26, 30–37, 44, 51
Montague, William Pepperell, 33
moral equivalent of war concept: and controlling energy in atomic age, 139–42; and energy crisis in 1970s, 172; and energy language, 13–14; and Green New Deal, 4; and James, 13, 48, 67
Morgan, Arthur, 103
Morgan, Thomas Hunt, 39
Morrison, Paul, 121
Motherwell, Robert, 150
Muir, John, 101
Mumford, Geddes, 143
Mumford, Lewis: criticism of, 150–51, 152–53; and culture, 82–83; and decline of civilization, 88; and energetic ethics, 19; and fascism, 150–51, 156; and Geddes, 93; international travel, 16; and libido, 84–87, 145; and new man concept, 142, 143–45; planning philosophy, 93–94; and psychic energy, 18, 81, 84; and RPAA, 85, 91; and Spengler, 88–89; and Tansley, 92, 93–94; and world unity, 162
mysticism: and energetic ethics, 156, 157–58, 161; interest in, 37, 142, 151, 156; and Osborn's energy laws, 39

Nadel, Alan, 139
Nagel, Ernest, 151
Napoleon I, 61
National Union for Social Justice, 121

natural resources: Appalachian Trail as, 97; and conservation movement, 23, 100–102; and Teddy Roosevelt, 23, 48–49
neurasthenia, 40–41, 44, 73, 196n119
New Age group, 112, 121
New Deal: and anti-Semitism, 120, 121; and energy language, 11, 109–10, 122–26, 134, 173; and Julian Huxley, 140, 161; and Olson, 149; and Paterson, 131, 132, 133; and replacing money with energy, 18, 131, 134
New Deal, Green, 4
New England Power Company, 96
new man concept, 142, 143–45, 168
Nichols, Louis B., 75
Niebuhr, Reinhold, 151
Noguchi, Isamu, 129
noosphere, 167
Nordau, Max, 40
Norris Village, TN, 125–26
Nye, David, 4, 25, 83

objectivism, 18. *See also* Rand, Ayn
oil, 3, 4
Olson, Charles, 19, 142, 149, 171
Omega point, 167
Ontario Hydro-Electric Commission, 96
Orage, A. R., 112
organic farming, 89, 90
orgone, 81, 86, 143, 155
Osborn, Henry Fairfield, 26, 38–40, 49
Ostwald, Wilhelm: admiration for Germany, 49–50; and energetics, 15, 30–32, 34–35, 36, 67–68, 109; and individualism, 47, 49; and monism, 15, 17; and sunspots, 67; and will, 17, 25–26, 49; and world unity, 166

paleotechnic phase of civilization, 83
paradox, energy: and climatic energy, 57, 80; defined, 14–15; and energetic ethics, 155–61; and infrastructure planning, 18, 84, 106; and psychic energy, 18, 84, 106; and replacing money with energy, 110, 134, 135; and will, 45, 52; and World War I, 11–12
Paris Climate Agreement, 2
Parrish Wayne W., 122–23
Paterson, Isabel, 18, 110, 131–35, 152, 173
Pearl, Raymond, 29–30
Pinchot, Gifford, 94

planning. *See* infrastructure planning; state planning
playground movement, 75
Political and Economic Planning (PEP), 140, 161
politics: and dynamic versus static societies, 133–34; and energetic ethics, 19, 139, 142–43, 161–65; and energy language, 3–4, 18, 109–10, 122–26, 134, 173; and replacing money with energy, 18, 109–10, 112, 115, 118–22; and strenuous life paradigm, 17, 24, 26, 47–49; and sunspots, 61, 63. *See also* communism; democracy; New Deal; reform movements
Pollock, Jackson, 19, 142, 150
pop culture, anxiety about, 153
positive environmentalism, 74–75
Pound, Ezra, 18, 121–22, 149
power (energy): abundance of in United States, 5, 6–7; development in United States, 25; effect on culture, 5, 25, 82–84, 95; and energy as term, 5; and energy consciousness, 3; and energy conservation in 1970s, 4, 171–72, 174; and energy policy, 3–4; and Giant Power, 82, 84, 94–99; and Green New Deal, 4
power utilities, 94–95, 96, 123–24. *See also* Giant Power; TVA (Tennessee Valley Authority)
pragmatic liberalism, 144
President's Commission on National Goals, 4
primitive/primordial energy: America as closer to primordial energy, 90, 100–102, 106; and civilizations, 146; Paterson on, 132; and psychic energy, 84, 87–91, 99–102; and Spengler, 146
progressive energy. *See* will
Progressive Era: and energy as positive, 3, 11
Progressive era: decline of Progressive Party, 50; and efficiency movement, 15; overview of, 17, 23–26; strenuous life paradigm, 17, 24, 26, 47–49. *See also* Roosevelt, Teddy; will
"Projective Verse" (Olson), 149
psyche, geographic and waterway metaphors for, 91–94
psychic energy: and culture, 18, 81–87, 97–99; and energy paradox, 18, 84, 106; geographic and waterway metaphors for, 91–94, 99, 104–6; and Giant Power, 82, 84, 94–99; and modernity critiques, 84, 87–91; and orgone, 81, 86; overview, 18, 81–84; as popular

psychic energy (continued)
  metaphor, 85, 105–6; as primitive/
  primordial, 84, 87–91, 99–102; and
  psychoanalysis, 18, 83, 84–87; and regional
  planning, 18, 83–84, 102–5, 173; and TVA, 18,
  83–84, 102–5; and world unity, 167
psychoanalysis: and energetic ethics, 141; and
  geographic metaphors for psyche, 91–94;
  and Julian Huxley, 161, 162, 166, 173; and
  libido, 85–87; and Marxism, 81, 86; and
  psychic energy, 18, 83, 84–87; and relative
  value of energy, 12

Rabinbach, Anson, 5, 6, 10
race: and civilization, 167; and climatic energy,
  17–18, 56–57, 65, 70–71, 73, 76, 78–79; and
  energy language, 16; and monism, 185n40;
  and world unity, 167
radio, 55–56, 58, 59–60, 127
Rand, Ayn, 131, 152–53
Rautenstrauch, Walter, 108, 113, 114
reform movements: and anti-Semitism, 109,
  120–22; and energy language, 109; and
  positive environmentalism, 74–75; and
  psychic energy, 89–91; and replacing money
  with energy, 18, 109–10, 112, 118–22. See also
  Technocracy movement
Regional Planning Association of America
  (RPAA): and folk revival movement, 91; and
  Giant Power, 94–99; international ties, 16;
  and Mumford, 85, 91; and Norris Village,
  126; origins of, 85; and psychic energy, 18,
  84, 173; and Spengler, 88; and TVA, 103
Reich, Wilhelm: geographic metaphors for
  energy, 92; and libido, 87, 92; and orgone, 81,
  83, 86, 143, 155; and psychic energy, 81, 84
relative value, energy as: and climatic energy,
  57; as debatable, 10–14, 17–19; and need for
  state control, 13–14; and quality versus quantity
  of energy, 13; and replacing money, 110, 118,
  126–28, 129, 132–33, 134; and US involvement
  in World War I debate, 1–2, 12; and will, 26.
  See also intrinsic/objective value, energy as
religion: and art, 145; and capitalism, 154; and
  Coughlin, 120–21; and mechanistic concept
  of energy, 29–30, 37; and monism, 32; and
  Mumford, 145; and replacing money with
  energy, 126; and revelation, 163; and

scientific humanism, 162–65, 166–68;
  sociolatry, 155; sun worship, 89; and Technocracy, 116, 126
Renaissance, 69–70
resources: and Giant Power's effect on culture,
  97–99; psychic energy as, 84, 96–102, 103;
  TVA as psychological resource, 103. See also
  natural resources
revelation, 163
Richards, R. O., 49
Rinkinen, Jenny, 174–75
Roosevelt, Eleanor, 126
Roosevelt, Franklin D.: and Coughlin, 121; and
  energy language, 18, 122–24, 130; and RPAA,
  103; and Technocracy, 11, 112, 115; and TVA,
  102. See also New Deal
Roosevelt, Teddy: and Brooks Adams, 43, 50, 51;
  and conservation of natural resources, 23,
  48–49; and energy language, 3, 16; and Loeb,
  23–24; mix of energy approaches, 48–49, 51;
  and Osborn, 39–40; and strenuous life
  paradigm, 17, 24, 26, 47–49, 50
Rosenberg, Harold, 149–50
Ross, Edward A., 49
Rowan, William, 58
Royce, Josiah, 34
RPAA. See Regional Planning Association of
  America (RPAA)
rural revival, 89, 90
Ruskin, John, 114, 132
rusting bank notes, 121

Sachs, Julius von, 26–27
saints, 46, 47
Salter, Patricia J., 155–56
Schopenhauer, Arthur, 32
scientific humanism, 162–65, 166–68, 169
scientific management, 11
Scott, Howard, 108, 109, 110–14, 119–20,
  127, 173
Seow, Victor, 8
sex: and climatic energy, 76, 79; and energetic
  ethics, 155–56, 157, 158–60
Shaw, George Bernard, 29, 37
Shove, Elizabeth, 174–75
Sinclair, Upton, 37
Slosson, E. E., 187n72
Sluga, Glenda, 167

Smith, Al, 61
Smith, Crosbie, 5, 33
Smith, Geddes, 84–86
Smuts, Jan Christiaan, 29, 165
Snyder, Carl, 29
Social Credit movement, 18, 109–10, 112, 118–20, 121
Social Credit Party (Alberta), 119
social Darwinism, 26, 38–44, 185n40
social degeneration and climatic energy, 17–18, 56, 72–73, 77–79. *See also* civilization
socialism: and Paterson, 132, 133; and will, 17, 37
social value of energy. *See* relative value, energy as
Soddy, Frederick, 109, 111–12, 113–14, 120, 121, 124
solar constant, 58, 64, 71. *See also* sunspots
solar credit, 118–19
solar cycles. *See* sunspots
Sorokin, Pitirim, 67–68
Soule, George, 128, 144
Soviet Union: and Chase, 95, 151; electrification in, 95; and Julian Huxley, 161; lauding of, 151–52; and totalitarianism, 150, 151, 152
Spencer, Herbert, 40, 45
Spengler, Oswald, 84, 88–89, 98, 146
state planning: and Brooks Adams, 43–44; and energetic ethics, 19; and monism, 32
states' rights and TVA, 105
Stein, Clarence, 85
Stetson, Harlan True, 59–60
Stokley, James, 59
Stormer, Eugene, 8
strenuous life paradigm, 17, 24, 26, 47–49
sun: and heliotropism, 26–27, 28, 30, 42; solar credit, 118–19; sunspots, 56, 57–68, 78, 88; sun worship, 89
"Sun, The" (Lawrence), 90
sunspots, 56, 57–68, 78, 88
superego, 87, 141, 199n27
Surén, Hans, 200n48
Sutter, Paul, 9
Swer, Gregory Morgan, 87
Szeman, Imre, 4, 5

Tansley, Arthur, 92–94
Taylor, Frederick Winslow, 11
Tchijevsky, Alexander, 55, 57, 60, 61, 66–67, 80

Technical Alliance, 112, 124
technic phase of civilization, 83
Technocracy movement: critiques of, 110, 114–15, 119, 120, 126–28, 130, 131–32; and replacing money with energy, 11, 18, 107, 109–18, 126–28; and yardstick metaphor, 122–23
technocratic management, 32. *See also* Technocracy movement
Teilhard de Chardin, Pierre, 166–67
Tennessee Valley Authority (TVA), 18, 83–84, 102–5, 123, 161–62
Tepoztlán, Mexico, 125
Thanatos (death drive), 12, 86
therblig standard, 130
thermodynamics, laws of: and dynamic versus static societies, 133; and energy language, 6, 7; and heat death, 25; and human work, 11; and neurasthenia, 41; and objective value of energy, 10; and replacing money with energy, 110–12, 120, 121; and social evolution, 37–44; and Technocracy, 110, 116; and wealth, 120; and will, 17, 33, 34
Third World Power Conference, 82–83
Tobey, Ronald, 83
Toews, John, 9
torpedo, heliotropic, 27
Torriti, Jacopo, 174–75
totalitarianism: and energetic ethics, 19, 141–42, 150–54, 156, 164, 168; fear of spread of, 18–19, 150–54, 156, 164; and infrastructure planning, 141; and Soviet Union, 150, 151, 152
tourism and wilderness, 100–101
Townsend Plan, 119
Toynbee, Arnold, 146–48, 151
trees and sunspots, 58, 59, 65, 78
tropics and climatic energy, 71, 73, 74, 76, 78–79
tropism, 26–27, 28, 30, 42
Trump, Donald, 2, 3
Turner, Frederick Jackson, 41, 100
TVA (Tennessee Valley Authority), 18, 83–84, 102–5, 123, 161–62
Tyndall, John, 33

UNESCO, 19, 139, 142, 143, 162, 165–68, 169
unified-energy-field concept, 66

United Nations Conference on International Organization, 139

United States: American energy as threatened, 3, 17–18, 25, 26, 56–57, 71, 73, 74, 77–79, 172; American exceptionalism, 73, 79; and dynamic versus static societies, 133; energy consciousness development in, 2–4, 172, 175; energy consumption, 2, 6–7, 8, 25; imperialism of, 43, 47–48, 71; and Paris Climate Agreement, 2; and World War I, 1–2, 6, 12, 24–25, 144; and World War II, 144. *See also* Americanness and American culture

Unwin, J. D., 158–61
usury, 120, 121, 160
utopia, 78–79, 116, 156–61

Veblen, Thorstein, 36–37, 112
Vernadsky, Vladimir, 165, 167
virginity and will, 42
vitalist concept of energy, 25, 29–30

war and sunspots, 61, 65, 66–67, 88
Washington, George, 43
waste of energy: and Chase, 108, 125, 173; and consumption, 36; and gender, 207n75
waterway metaphors for psychic energy, 91–94, 104–6
Watts, Sarah, 47
wealth: energy as, 112; as form of energy, 132; and individualism, 35, 36; and laws of thermodynamics, 120; versus money, 112; and replacing money with energy, 124–25
Wells, George Philip, 162
Wells, H. G., 162, 165
Wellum, Caleb, 4
Wheeler, Raymond Holder, 64–66
Whitaker, Charles, 85
Whiteness: as American, 56, 57; and climatic energy, 17–18, 56–57, 70–71, 78–79
Wiener, Norbert, 173–74
wilderness, 99, 100–102
Wilderness Act, 101
Wilderness Society, 101–2
Wilkinson, A. E. A., 33

will: and civilization, 146; and democracy, 17, 25, 43–44, 50–51; dynamogenic will, 46–47; and energetic ethics, 139, 140, 146; and energy paradox, 45, 52; and evolution, 37–44; and individualism, 17, 25, 34, 35–36, 44, 46, 47–48, 49, 50–51; and James, 17, 26, 44–48, 49, 51–52; and mechanistic concept of energy, 25, 26–30, 35–37, 51; and monism, 17, 25–26, 30–37, 44, 51; and neurasthenia, 40–41; overview of, 17, 23–26; and social Darwinism, 26, 38–44; and strenuous life paradigm, 17, 24, 26, 47–49; and sunspots, 64; and Technocracy, 130; and world unity, 139–40

Wilson, Woodrow, 50
Wise, M. Norton, 33
women: and replacing money with energy, 117; sexuality and climatic energy, 76, 79
Woodman, E., 78
work: and energy language, 6, 7, 174–75; quality versus quantity of, 13; quantifying energy of, 11, 13, 34, 126–27, 173; and replacing money with energy, 126–27; and time-motion studies, 11
world unity. *See* internationalism and world unity
World War I: and Brooks Adams, 51; debate over US involvement, 1–2, 6, 12, 24–25, 144; and hyperinflation, 113; and Loeb, 183n3; and Teddy Roosevelt, 24–25
World War II, 102, 141, 144
Wright, Henry, 85, 103

yardstick, energy as: coinage of phrase, 104; and decline of civilization, 79; and democracy, 81; and energetic ethics, 143, 164–65, 168, 169; and energy as objective value, 11, 109; and Fuller, 128–31; and individual, 143, 164–65; political use of, 122–26; and replacing money with energy, 113–18; and Technocracy, 122–23; and TVA, 104, 123
Young's cycle, 55

Zimmermann, E. W., 125

Explore other books from HOPKINS PRESS

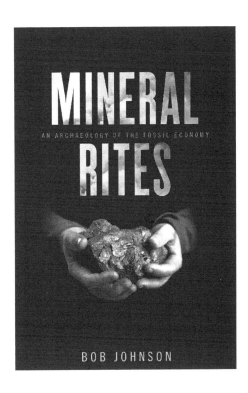

"This book is indispensable for understanding the centrality of fossil fuel popular culture to the shift toward neoliberalism and the New Right."

—Cara New Daggett,
author of *The Birth of Energy*

"A highly original, imaginative book that offers an alternative history of modernity."

—Natasha Zaretsky,
author of *Radiation Nation*

JOHNS HOPKINS UNIVERSITY PRESS | PRESS.JHU.EDU